THE

CONSTRUCTION

OF A

THERMOMETER

THE CONSTRUCTION OF A THERMOMETER

by

JAMES SIX F.R.S.

Prefaced by an Account of his Life & Works and the Use of his Thermometer over Two Hundred Years by Jill Austin & Anita McConnell

NIMBUS BOOKS

ACKNOWLEDGEMENT

We are grateful for the contributions of Mr D. Harrington, historian, of Canterbury, to Mr H. A. Harvey who allowed us to see the Archives of King's College, London, and to Mr V. K. Chew of the Science Museum, London, for advice and guidance.

THE CONSTRUCTION AND USE OF A THERMOMETER
was originally published at Maidstone in 1794.
This edition is printed in 1980 by Nimbus Books
with additional material.

84 Sydney Road London N10

ISBN 0 9507036 0 5

Printed and Bound in Great Britain by
The Scolar Press, Ilkley, Yorkshire.

CONTENTS

FOREWORD

In 1782 the Royal Society published an account of a thermometer invented by James Six in 1780*. This remarkable instrument recorded both extremes of temperature in the absence of the observer while being robust and simple to use. Its invention was considered sufficiently important for James Six to be elected a Fellow of the Royal Society in 1792. Since then the Six's thermometer has played a worthy part in oceanography and meteorology and its basic design has changed little in two hundred years. Today more accurate instruments are available for professional work but the Six's thermometer remains a popular and reliable domestic instrument, to be found in many homes and gardens.

Until recently none of Six's original thermometers were known to survive with the possible exception of one at the Museum for the History of Science at Oxford. Research now shows that the Science Museum, London, possesses twenty-six instruments, designed for use on land and at sea†. Their existence was not previously recognised since they form part of the King George III Collection of Philosophical Instruments, on loan from King's College, London.

Despite the widespread use of the instrument that bears his name, James Six is a long-neglected FRS who is not even listed in the Dictionary of National Biography. This volume, printed for the bicentenary of the thermometer, is an attempt to gather together and survey all Six's work. His book *The Construction and use of a Thermometer* (1794), his correspondence with *Gentleman's Magazine* and the American Philosophical Society, and a manuscript letter, are reproduced as they are not readily available. All are written with charm and simplicity and provide a rare insight into eighteenth-century techniques of instrument construction.

*SIX, J. An account of an improved thermometer. *Phil. Trans.* *72*, pp.72–81.
†AUSTIN, J. F. & McCONNELL, A. James Six, FRS: Two hundred years of the Six's self-registering thermometer. *Notes & Records Roy. Soc.* *35* (1) (1980).

THE LIFE AND WORK OF JAMES SIX

1731–1793

James Six was descended from one of the refugee Walloon families which had settled in Canterbury during the reign of Elizabeth I. Many of these families were silk weavers and they were allowed to establish their trade in that city. Charles II later granted them a Charter of Incorporation which brought into existence the Company of Silk Weavers in the City of Canterbury. The name of the first master was James Six. The Canterbury silk trade reached its peak towards the end of the seventeenth century and subsequently had to compete against cheap imported silks from India and Persia. The invention of 'Canterbury Muslin' by John Calloway in the closing years of the eighteenth century only temporarily stemmed the flood of imported goods. As a young man Six was brought up in the family silk business but declining trade prompted him to retire early and devote himself to natural philosophy. There is no evidence that he ever received any formal scientific training.

All accounts of Six describe him as a modest, unassuming man of great integrity. He displayed the catholic range of interests typical of an eighteenth-century intellectual, concerning himself with painting, astronomy, electrical machines and thermometry. His formal published works so far located deal only with the construction and use of his thermometers.

'An account of an improved thermometer' was read to the Royal Society on 28 February 1782. This paper gave details for constructing a thermometer for use in air. The invention was obviously no academic exercise but was motivated by Six's interest in meteorology. Subsequently he quoted observations from his thermometers in two papers which were read to the Royal Society in 1784 and 1788. For this, thermometers were set at intervals up the tower of Canterbury Cathedral, and on top of a local hill. Daily reading were taken from all the instruments for many months. The papers were a summary of his interesting and detailed observations on the variation of temperature with height under differing weather conditions.

Six discovered, without much surprise, that during the day temperature almost always decreased with height. However, he made the important and unexpected discovery that the vertical gradient usually reversed at night, so that it was coldest near the ground. Six noticed this effect was particularly marked on clear nights and deemed the finding 'extraordinary'. Not until a century later was it realised that the night-time inversion was due to intense radiational cooling at the earth's surface.

In 1794 *Construction and use of a Thermometer* was published posthumously. This book summarised Six's published papers on thermometers and also contained a chapter on the adaptation of the self-registering thermometer for deep-sea use.

Six corresponded regularly with *Gentleman's Magazine* and in these letters showed himself to be a competent astronomer with a thorough knowledge of the night sky. In 1770 he described a comet, later known as Lexell's Comet, during its approach to and from the sun. His observations proved invaluable to the American astronomer David Rittenhouse, as they fitted almost exactly Rittenhouse's calculated orbit for the comet*. Six recorded the appearance of two further comets in 1771 and 1781. In 1783 he corresponded on the anomalous weather of that July and in 1784 sent in his observations of the new 'Georgian Star', the planet Uranus, which had been discovered in 1781 by William Herschel. His later observations of Uranus were communicated to the American Philosophical Society by Rittenhouse in 1786†. Six must have possessed a fairly elaborate telescope with interchangeable eyepieces as he used a variety of magnifications for his observations.

Six visited the Royal Society about once a year after the publication of his 1782 paper. On 23 June 1791 he was proposed for election as a Fellow in recognition of the invention of his thermometer. His proposers were Nevil Maskelyne, Francis Wollaston, Alexander Aubert, John Smeaton and Samuel Simmons. A notice was posted in the usual fashion, being read ten times in the course of the next six months. A ballot was taken on 19 January 1792 and Six was duly elected. In the remaining year of his life he neither proposed any other persons for Fellowship nor served on any of the Society's committees.

Six was also a foreign member of the American Philosophical Society which had in the past competed vigorously with the American Society for Promoting and Propagating Useful Knowledge; as a result both bodies regularly elected non-resident members. The two societies amalgamated in 1768 but continued to recruit abroad. This probably explains why James Six '. . . received very unexpectedly from America a certificate of his inrollment as a member of the American Philosophical Society . . .'

Six's private life was the tragic but then familiar tale of a parent who outlived most of his children. On 21 May 1754 he married Mary Riquebourg of St Mildred's parish, Canterbury; of their five children, three sons died in infancy‡ and the remaining son, James Six junior, grew up to be an outstanding scholar§. He attended the King's School, Canterbury and his great friends there were William Friend and Will Tylden, all three went up to Cambridge University together||. Friend was well-known to the Six family and possessed some of the father's thermometers. In 1785 James Six senior sent him some thermometer indexes along

*HINDLE, Brooke. *David Rittenhouse*. Princeton (1964), p.76.

†RITTENHOUSE, D. Astronomical observations by Mr James Six, of Canterbury. *Trans. Amer. Phil. Soc. 2* (1786), p.260.

‡COWPER, J. M. *Memorial inscriptions in the Church and Churchyard of Holy Cross, Westgate*. Canterbury (1888), pp.46 and 49.

§*Alumni Cantabrigenses* II, p.524, Cambridge (1953).

||KNIGHT, F. *University rebel, the life of William Friend*. London (1971), p.18.

with detailed instructions for inserting them into the tubes*. In the summer of 1786 young James set off with his two friends to visit Paris, Switzerland and Italy. It was a fateful journey for he never returned, dying of fever in Rome at the age of twenty-nine. The surviving daughter, Mary, married George May, an 'opulent brewer'. Through this line of the family Six's original thermometers were preserved and handed down.

James Six was a devout Christian throughout his life. With Robert Deane† he founded and supported a Sunday School in the parish of Holy Cross. The School was for the religious instruction of 'poor innocents' who were also taught reading and singing‡. After his son's death in Rome, Six placed a memorial to him in the Holy Cross parish church at Westgate, Canterbury and helped to buy an organ at the same church. James Six died on 25 August 1793 at the age of sixty-two. His wife Mary died on 19 March 1801 and they are buried together in a vault in Westgate Church†.

*This letter is reproduced on p.22.
† COWPER, J. M. Op. cit., pp.3 and 2.
‡ HORNE, G. *Sunday Schools recommended in a sermon preached at the parish church of St Alphage, Canterbury, on 18th Dec. 1785.* Oxford (1786), pp.33–6.

THE SURVIVING SIX'S THERMOMETERS IN THE SCIENCE MUSEUM, LONDON

The Science Museum acquired Six's thermometers by chance, without appreciating their origins and importance. James Six's will bequeathed all his goods to his wife Mary and, upon her death, to his daughter Mary May. One of her descendants presented the thermometers to King's College, London, in the mid-nineteenth-century and they were incorporated into the Wheatstone Collection. This was housed in the same room as the King George III Collection of Philosophical Instruments and inevitably the two collections were not kept strictly apart. The thermometers' provenance was apparently forgotten and they were transferred to the Science Museum in 1926 as part of the King George III Collection*.

Of the twenty-six thermometers, twelve are for use in air. The design is illustrated and described in Chapter I of Six's book, so it remains to note that Six's innovation was to place the slender glass-tailed indexes *within* the thermometer tube, a feature that has been copied in many later instruments. The twelve thermometers are complete, except for one which lacks the central reservoir. Their construction is crude, with rough wooden frames and irregularly-blown glass (see Figure 1). All the instruments save two are of similar dimensions, being about 350 mm long. One has an exceptional length of 565 mm; the other is 170 mm, with a double reservoir (see Figure 2). Mean external diameter of the alcohol reservoirs is 7.5 mm and that of the glass tubing 3.9 mm.

*CHALDECOTT, J. *Handbook of the King George III Collection of Scientific Instruments.* (London (1951).

It has not been possible to measure the internal diameters of the tubes.

Only five of the thermometers are calibrated and this is done by pencil marks. The ranges vary; the lowest temperature marked on a minimum arm is —40°F and the highest on a maximum arm is 110°F. The alcohol has long since evaporated in the maximum arms but it has been possible to check some of the 'minimum' calibrations and these proved remarkably accurate compared with readings from a modern mercury-in-glass thermometer, differences not exceeding 1 °F. None of the scales is linearly spaced. This is because Six was aware of the irregularities in the bore and so calibrated his instruments step-by-step against a standard thermometer. This was obviously preferable to the then usual method of taking two fixed temperatures and interpolating regular graduations. Most of the thermometers contain their original indexes. Six instructed his readers 'Provide yourself with some of the finest sewing-needles, nip off a third of their length from the end which has the eye and keep the remainder for use'. The needles were then mounted in small glass tubes, the ends sealed with 'bugles' and long glass tails drawn out. Bugles are small buttons of black glass which make the end of the index more visible. It has been possible to extract one of the indexes and examine it under a microscope; its construction matches Six's description.

Six considered that his meteorological thermometer would be impractical for marine use because its indexes would be dislodged by the buffeting of the waves. He therefore substituted indexes which consisted of '. . . . a piece of needle or steel wire, an inch long, hermetically sealed in a very thin glass tube, just large enough to contain it.' Before use, each needle was brought to rest on the mercury surface. As the mercury descended in one arm, so did the needle resting on its surface. However, when the mercury rose again, it flowed past the needle which remained to mark the maximum or minimum temperature reached. Thus the indexes in the sea thermometer acted in the opposite sense to those in the air thermometer, in Six's words, '. . . . accomplishing by descending and remaining at their lowest point of descent, what the others do by ascending and remaining at their highest point.'

There are fourteen marine thermometers in the Science Museum collection, all of them complete. One is shown in Figure 1. Slightly smaller than the air thermometers, their average length is 330 mm. The glass is noticeably thicker to withstand the anticipated deep-sea pressure and some of the instruments have 'thick' pencilled on them. The mean external diameter of the alcohol reservoir is 10.0 mm and that of the tubing is 4.3 mm. Three of the thermometers are calibrated and they show the rather smaller ranges appropriate for sea water. It seems unlikely that any of the instruments were ever submerged for the wooden frames show no sign of water damage. One of the thermometers retains the brass capping which Six considered essential for keeping seawater out of the open limb.

This design should have worked very well because, once the

needles were immersed in the mercury, surface tension pressed them so strongly to the side of the tube that it was impossible to move them, even by shaking or using a powerful magnet. To retrieve the minimum index it was necessary to cool the instrument below the minimum temperature to which it had been exposed. Six recommended applying strong 'spirit of wine' to the outside of the alcohol reservoir, to cool it by evaporation. When not in use, the indexes were stored well clear of the mercury in two small reservoirs at the top of the instrument. These reservoirs are shown in detail in Figure 3 and in Plate III of Six's book. However, resetting the indexes must have been a problem because the marine thermometer was never developed. The original meteorological thermometer had to serve both on land and at sea.

DEVELOPMENT AND USE OF THE SIX THERMOMETER ON LAND AND AT SEA

Six did not patent his thermometer but instrument-makers were slow to copy the design, almost certainly due to the difficulties of construction. In 1787 an anonymous letter in *Gentleman's Magazine* stated that two Six's thermometers had been despatched abroad, one to Paris and the other to Italy. In Tuscany Signor Antonio Matteuci had made several copies of the newly-arrived thermometer, presenting one to the Duke of Tuscany. The writer of the letter hoped that this 'may stimulate some of our mathematical instrument makers to make thermometers on this principle'. A further letter in 1789 compared observations from the thermometer in Italy with one in Canterbury.

It seems that the Six thermometer first became commercially available in the early 1790s. The posthumous inventory of Dr Priestley's laboratory in 1791 listed a thermometer 'of Mr Six's Construction' and also a barometer 'constructed in the manner of Mr Six's thermometer'. These two items were valued at two and three guineas respectively by Messrs Nairne & Blunt*. A Six thermometer was made about this time in America by Joseph Gatty, an Italian instrument-maker resident in Philadelphia. The thermometer was for George Washington, who kept a regular weather diary. It is preserved in Washington's house at Mount Vernon, while a copy is on display in the Smithsonian Museum†.

Few systematic meteorological observations were taken at the end of the eighteenth century. However, in 1794 the Royal Society adopted the Six thermometer for recording maximum and minimum temperatures and the Royal Observatory at Kew followed suit in 1798‡. It rapidly became the practice to average the maximum and minimum readings and call this quantity the

*BOLTON, H. C. *Scientific correspondence of Joseph Priestley*. New York (1892), p.225.

†KNOWLES MIDDLETON, W. E. *Catalogue of meteorological instruments in the (Smithsonian) Museum of History and Technology*. Washington (1969), pp.46–7.

‡Meteorological Register for Kew Observatory, vol.2. Archives of King's College, London.

mean daily temperature, which was then used in the compilation of climate statistics. From 1807 Luke Howard used a Six thermometer in this fashion to record average daily temperatures for his study on the climate of London*.

The inherent failing of the Six thermometer was the tendency of the alcohol to 'creep' past the mercury column and displace it relative to the scale. The accuracy of the thermometer was impaired by low standards of workmanship and departures from Six's original design. The proportions of the bulb and tube had changed by the 1790s, judging from the replica of Washington's thermometer illustrated in the Smithsonian Catalogue. Descriptions of the thermometer from the early nineteenth century reveal that the tube was usually sealed and the overall size was reduced; this in turn demanded smaller indexes. The long fragile glass tails were replaced by several hairs or bristles, tied to the upper bugle and laying alongside the index itself. Immersed in alcohol, the natural spring of these bristles eventually weakened so that the indexes became loose. Sometimes the liquids were not sufficiently cooled before filling the thermometer so that bubbles of air formed in the alcohol at low temperatures. If the tube was sealed with insufficient air pressure in the space above the liquids then the columns of alcohol and mercury separated at low temperatures. In 1833 Luke Howard complained bitterly about the changes which had been made to the Six thermometer, saying that it 'had been degraded in the hand of successive workmen'*. Daniell in 1845 found its construction often 'shamefully deficient', remarking that little care was taken by instrument-makers to achieve a tube of even bore†.

Bearing these details in mind, what is the likely date of the Six thermometer at Oxford, which is mounted in a frame identical to that illustrated in Six's book? The tube is sealed and the indexes have hair springs so that it is likely to have been constructed after Six's lifetime, probably during the early nineteenth century. There is a Six's thermometer with glass tails on the indexes in the Copernicus Museum in Rome, but this is the only example we have seen apart from Six's original instruments that reproduces the original design. Its origins are obscure, but it was probably obtained for the Observatory at the Collegio Romano.

Thus for accurate meteorological work the Six's thermometer declined in popularity throughout the nineteenth century. It was gradually replaced by the Rutherford separate maximum and minimum thermometers‡ which used small glass cones as indexes and hence had to be mounted strictly horizontally. This delicate instrument was acceptable on land, but the Six's thermometer was still better suited to observations in the deep sea or on balloon ascents§.

*HOWARD, L. *The climate of London*, vol.1, London (1833), p.xii.
†DANIELL, J. F. Elements of meteorology. London (1845), vol.2, pp.269–70.
‡RUTHERFORD, D. A description of an improved thermometer. *Trans. Roy. Soc. Edinburgh 3* (1794), pp.247–9.
§de FONVIELLE, W. *Les ballons-sondes de MM Hermite et Besançon.* Paris (1898).

Despite its shortcomings, one can see from instrument-makers catalogues of the late nineteenth century that the Six thermometer was popular for horticultural and domestic use, where accuracy was not paramount. It was offered for sale in a variety of forms suited to indoor and outdoor use, in the guise of a 'scientific instrument' and as 'furniture'. Two- and three-limbed forms were made; proportions of the bulb and tube varied considerably, but gradually the U-shaped tube with close-fitting indexes became the accepted type and is the familiar thermometer of today.

There is no evidence that Six's marine thermometer was ever produced commercially. In 1798 General Aboville suggested that the air thermometer could be converted to underwater use by sealing the tube and adding a second mercury sector, whose weight would replace the restoring affect of atmospheric pressure*. This latter modification proved unnecessary.

The first known use of the Six thermometer at sea was in 1803–6 during the circumnavigation of the world by the Russian ships *Neva* and *Nadezhda*. We do not know the precise form of their thermometer, but on occasion the results were at variance with those obtained on deck from samples recovered with a water-bottle†.

William Scoresby, whaling captain and Arctic scientist, gives one of the earliest reports of its use on a British ship‡. From May 1811 onwards Scoresby sent down a Six thermometer (lent to him by Sir Joseph Banks, PRS) in his water-sampler, confirming his belief that in the Greenland Sea the coldest water lay at the surface, contrary to observations in more temperate latitudes.

The Admiralty was concerned with the possible existence of a Northwest Passage linking the Atlantic and Pacific Oceans. In 1818 two expeditions carrying versions of the Six thermometer were sent to the Arctic to test the water in various inlets. It was hoped that they would identify a strong current of saline water, less cold than that at the surface, as indicative of a channel leading to an open polar sea. However, the results were inconclusive.

Accounts of the use of these thermometers are found in the narratives of the several Arctic voyages taken by John Ross and Edward Parry§; we also have Sabine's account of his trials to determine the effects of pressure, undertaken during the same period‖ and it is apparent that the unknown pressure factor and the tendency of the indexes to shift under rough handling caused the readings of the thermometers to be viewed with suspicion by some observers, though not all.

Attempting to prevent deformation of the bulb under pressure,

*ABOVILLE. Changement proposé au thermomètre de l'Anglais Six . . . (*Paris*) *J. Mines 9* (1798–9), pp.75–6.

†von KRUSENSTERN, A. J. *Voyage around the world* . . . London (1813).

‡SCORESBY, W. *Account of the Arctic regions* (2 vols), Edinburgh (1820), pp.184–7.

§ROSS, J. *A voyage of discovery* . . . London (1819).
PARRY, W. E. *Journal of a voyage* . . . London (1821); *Journal of a second voyage* . . . London (1824), *Journal of a third voyage* . . . London (1826).

‖SABINE, E. On the temperature at considerable depths of the Caribbean Sea. *Phil. Trans. 113* (1823), pp.206–10.

since no correction could be arrived at for this factor, James Glaisher and Admiral Fitzroy approached the firm of Negretti & Zambra for a solution to the problem, and in 1857 the first pressure-protected thermometer was produced*. The thermometer bulb was surrounded by an outer glass sheath and the enclosed evacuated space partially filled with mercury. The mercury provided good thermal conductivity while the vacuum allowed for the flexure of the outer sheath under compression. Fifty thermometers were made to this pattern and supplied to Royal Navy vessels. These instruments were some eighteen inches long, mounted on wooden backs, and they were sent down in valved copper cases of the type invented by Robert Jamieson in 1820†.

By the second half of the nineteenth century the Navy's interest in Arctic explorations was replaced by the search for new forms of marine life and by sounding surveys for laying telegraph cables. In 1869 zoologists working under the aegis of the Royal Society drew the attention of its Vice-President Dr W. A. Miller to the want of a thermometer capable of measuring accurately the various deep-water masses in which they were dredging. By this date the firm of Casella were making a Six thermometer of conventional three-limbed form, but only eight inches long and mounted on ebonite. This was known as the Hydrographic Office (HO Pattern) thermometer. Miller took this instrument as the basis for the famous Miller-Casella thermometer‡, apparently in ignorance of Negretti's earlier design. Constructed by Casella after exhaustive pressure-tank tests, it had a two-limbed tube whose bulb was protected by an outer glass sheath, as in Negretti's design, but the space between bulb and sheath was partially filled with alcohol and its vapour. It was mounted on an ebonite block of the same size as the HO thermometer and was sent down in a perforated copper case. Relative sizes and forms of the Negretti and Miller designs are shown in Figure 4.

Negretti & Zambra objected strongly to this copying of their design but they had neglected to patent it and had no grounds for legal action. Within a short time the Miller-Casella was being widely advertised both at trade exhibitions and in other instrument-makers' catalogues. It seems certain that besides Casella and Negretti & Zambra, notable makers such as Elliott, Pastorelli and Hicks constructed it. The host of smaller firms who offered the Miller-Casella for sale under their own names had probably purchased their stock wholesale from these larger makers.

Interest in marine biology culminated in the voyage of HMS *Challenger* which cirumnavigated the globe in 1872–6. All regular temperature measurements were made by Miller-Casella thermometers in which the thermometric fluid was a mixture of creosote and alcohol, to prolong the spring quality of the index bristles (a custom probably introduced around mid-century). Several

*GREAT BRITAIN BOARD OF TRADE. *Meteorological papers* No.1 (1857), p.55.

†JAMIESON, R. Description of the marine thermometer case . . . *Phil. Mag. 57* (1820), pp.294–300.

‡MILLER, W. A. Note upon a self-registering thermometer . . . *Proc. Roy. Soc. 17* (1869), pp.482–6.

thermometers were normally lashed to the sounding line a fixed distance apart to yield a serial obervation. The instruments proved less than ideal however, when the vessel entered Antarctic waters with their characteristically complex temperature profile. During 1874 some examples of the newly-invented Negretti & Zambra deep-sea reversing thermometer were sent out to the ship. Their performance on test was promising but the results were often at variance with those from the Miller-Casella thermometers.

After *Challenger*'s return her thermometers were subjected to exhaustive tests by Professor Tait* and it became clear that all were subject to pressure error and had been so crudely made as to be incapable of resolving the fine detail essential for zoological research. By this time the deep-sea reversing thermometer was proving itself to be a very accurate instrument and it soon came to replace the Miller-Casella version of the Six thermometer.

*TAIT, P. G. The pressure error of the *Challenger* thermometers. *Challenger Report* Narrative *2* Appendix A (1881).

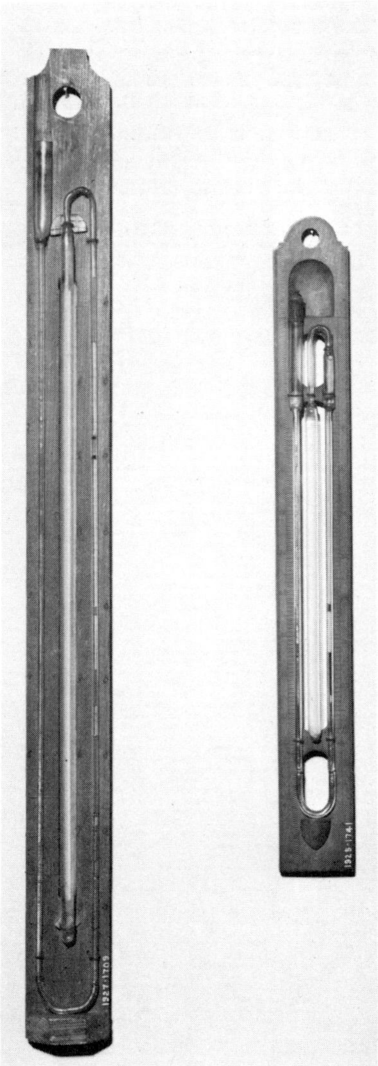

Figure 1
Original thermometers: left, for meteorological use; right: for deep-sea use.

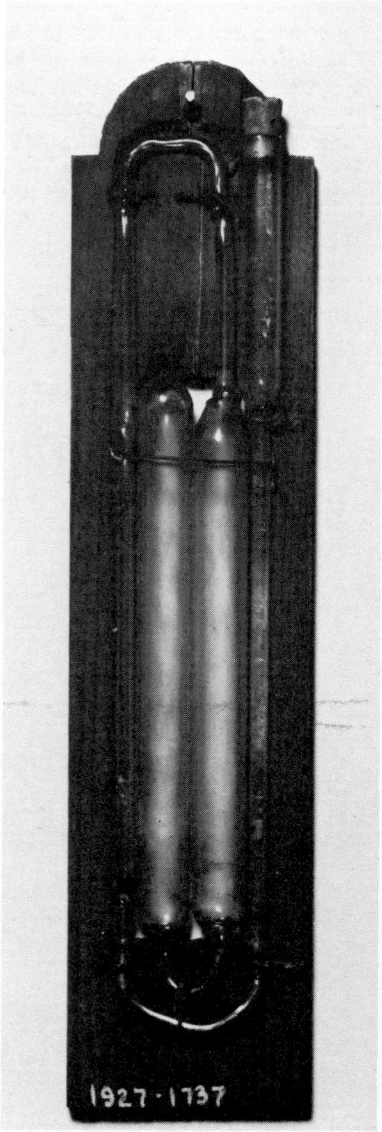

Figure 2
Compact four-limbed meteorological original thermometer.

Photo Science Museum London

Crown copyright Science Museum London

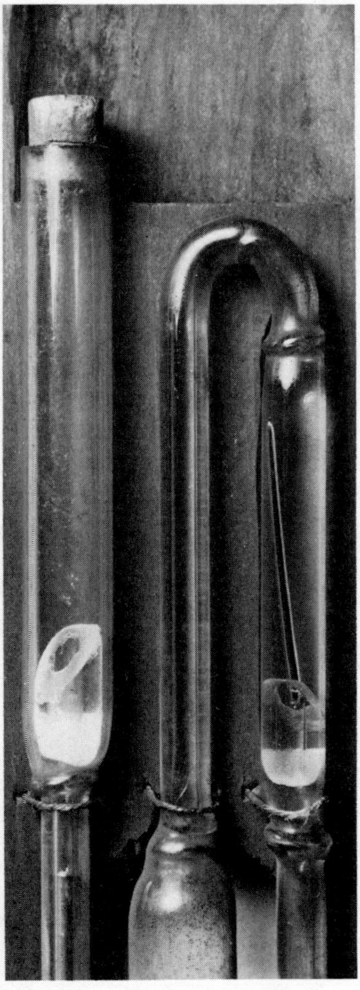

Figure 3
Detail of the storage chambers and index of an original deep-sea thermometer.

Photo Science Museum London

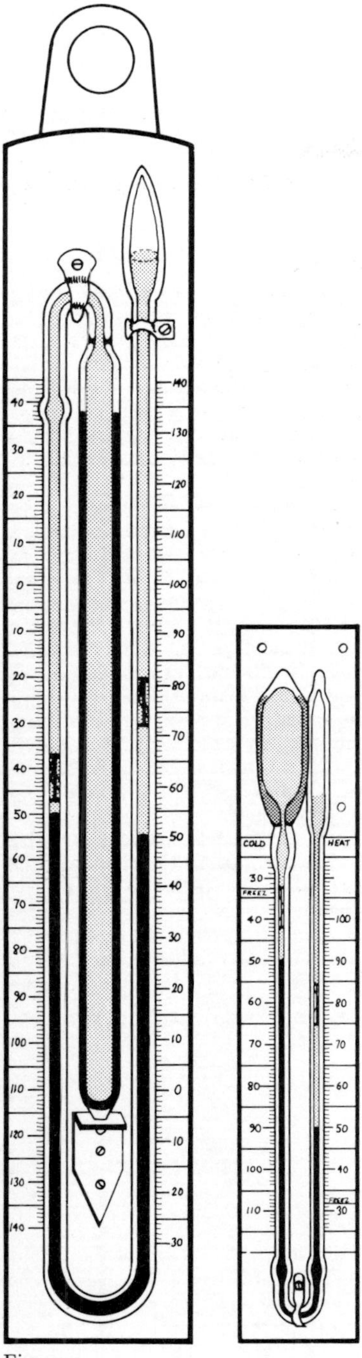

Figure 4
Construction and relative sizes of pressure-protected thermometers: left, by Negretti & Zambra, 1857; right, by Miller-Casella, 1869. Stipple represents alcohol and black represents mercury.

Photo Science Museum London

GENTLEMAN'S MAGAZINE *40* (1770), pp.319–20.

Obſervations on the Appearance of the late Comet.

Mr. URBAN,

AS the Comet which we lately ſaw in its approach to the ſun, is no longer viſible, and is not expected to appear again, and as from the unfavourableneſs of the weather, it was little noticed, (though the rapidity of its apparent motion was very remarkable), a ſhort account of it may be agreeable to ſome of your readers.

From Monſieur Meſſier's letter (*ſee occurrences*) it appears, that its motion from the 16th to the 21ſt of June, was not a degree and half; from that time to Mr. Dunn's obſervations on the 27th, it was only fourteen degrees; but in the ſix following days its velocity was ſo ſurpriſingly increaſed, as to paſs over 133 degrees in that ſpace of time; and from its apparent ſize and brightneſs increaſing in the ſame ſudden proportion, it ſhould ſeem that it paſſed very near the earth. Dr. Long * tells us, that from obſervations made on a number of Comets, " the greateſt apparent velocity any where mentioned, is that of the Comet in 1472, obſerved by Regio-Montanus to have gone through 40 degrees of a great circle in 24 hours;" but this laſt ſeems rather to exceed even that, for it moved 42 degrees in the ſame time, viz. from Saturday June 30, to Sunday-night July 1; which, ſuppoſing it no farther from us than the planet Venus at her neareſt approach, amounts to near twenty millions of miles.

Before I could get any inſtrument to take its place on Saturday evening, the clouds robbed me of the ſight of it, but the ſtars I ſaw near it, M. Meſſier's, Mr. Dunn's, and my own ſubſequent obſervations, inſtrumentally taken, confirm me in the opinion I had of its ſituation that night. For thoſe who would ſee its apparent courſe, on the celeſtial globe, the following obſervations will be a guide:

	R. Aſc.	S. Decl.
Between June 15 and 16, M. Meſſiers	272° 57′	—15° 45′
June 27, ½ paſt 11 at night, Mr. Dunn	275° 0	— 3° 0
		N. Decl.
June 30, at half paſt ten my own	280° 0	—35° 45′
July 1, Ditto at ditto	311° 40	—75° 0
Between 3d and 4th, at ½ paſt 12	86° 0	—49 0

Canterbury,
July 19, 1770. JAMES SIX.

* Long's Aſtronomy, vol. ii. p. 550.

GENTLEMAN'S MAGAZINE *40* (1770), pp.352–3.

Obſervations on the Appearance of the late Comet.

Mr URBAN,

SINCE my obſervation of the Comet publiſhed in your laſt, I have had the unexpected pleaſure of ſeeing it again, in its aſcent from the Sun towards its aphelion, and though it was not viſible to the naked eye, I ſaw it very diſtinctly with a teleſcope magnifying 25 times, when it appeared much like the *nebula* in *Andromeda's Girdle*, only a little ſmaller and brighter. On the 22d of Auguſt, at half paſt two in the morning, its place was in a line between δ and q in *Gemini*, on Senex's 17 Inch Globe, 1 deg. 18 min. from the firſt, and 18 min. from the laſt, of thoſe ſtars; it had 106 deg. 20 min. right aſcention, and 21 deg. north declination. On the 24th it appeared between p and r, having only increaſed its longitude 1 deg. 15 min. the latitude continued almoſt the ſame. On the 26th its longitude was increaſed 1 deg.

15 min. more, its latitude very little, not more than 5 min. from the first observation. Its apparent motion at present, nearly parallel to the ecliptic, is probably nothing more than its parallax occasioned by the earth's annual motion, which, by comparing the arc it made in passing us with its present situation, seems very considerable.

As the extraordinary apparent motion, and, of consequence, near approach of this Comet, in passing us, was very remarkable, so likewise its figure and appearance deserve our attention. On the 1st of July, the nucleus appeared as large as the planet *Jupiter*, surrounded with a coma of silver light, the brightest part of which was as large as the moon's orb, and, gradually diminishing its splendor, continued visible 3 or 4 deg. at least from the nucleus, and, as near as I could judge, (from the short time I viewed it) equal on every side. The remark I would make on this phænomenon, is, "that as the generality of astronomers at this time, acknowledge but one kind of Comets, and account for the different appearances of them, from the diversity of their situation with respect to the earth," * this last Comet seems to have proved the contrary: for, though it is certain, that some extend their coma, or tail, to a vast distance in opposition to the Sun, it seems as certain, that this had no such tail; for the direction of its course, and near approach to us, were such, that, I think, we must have seen it, had there been any: neither do the curious conjectures of the learned, concerning the nature of that light, as yet agree at all with the equal rays of this coma. From which, I presume, that we are still in the dark, in respect to the nature of Comets, but that an amazing and beautiful variety appears in those magnificent works of the Almighty, the heavenly bodies.

Canterbury,
Aug. 26, 1770.

JAMES SIX.

GENTLEMAN'S MAGAZINE *41* (1771), pp.28–9.

Six's Account of a late Comet.

MR. URBAN,

THE following particulars may be agreeable to some of your readers. On the evening of the 9th Inst. after viewing the Planet Saturn, I saw something near the Equator, which, at first, I thought, resembled a Nebula, but on examination I found it to be a Comet. Though it appeared faint, the nucleus seemed very distinct, surrounded with a coma, which, extending towards the South East, formed a tail of about five or six degrees apparent length. At first, by its situation in the Heavens, and the brightness of the nucleus, I had reason to think the Comet was descending towards the Sun; and therefore hoped soon to have a nearer view of it; but from observations on the succeeding nights, I perceived its apparent motion decrease, and it gradually became less distinct: and a few nights after, it was not to be seen at all; from whence it appeared to be returning from the Sun towards its aphelion. Its apparent course was North West, moving in three days 29 degrees, 30 minutes, which was the more considerable, as it seemed at a great distance from us. Tracing its course on the Globe two or three days back, we shall find it had then risen about midnight, and, besides, had so much South declination as to bring it low in our Horizon, which will, in some measure, account for its not having been seen here soon after its return from the Sun; but in Latitudes to the Southward of us, it is likely to have been more conspicuous. In its way towards the Sun, the Comet would probably have been visible in the Northern Hemisphere, had it not been for the long continuance of cloudy weather about that time. However, I had the pleasure of observing it several nights in its return, and found its place as follows:

ho.	rt. aſc. ° ′	n. dec. ° ′
1771, Jan. 9 at 11 *p. m.*	130 40	0 25
—— 10 at do.	121 20	5 20
—— 11 at 8	113 20	10 15
—— 12 at 9	104 35	14 45

Was it not already known that there are a conſiderable numb r of comets, it might ſeem ſtrange that we have been viſited by three in little more than thirteen months. Tables with 49 different appearances of this kind are now publiſhed, 20 of which have been obſerved ſince the year 1700, and tho' ſome of theſe, without doubt, deſcribe the return of the ſame Comet; yet probably the greater part are all different ones. It is likewiſe poſſible that many ſmall Comets paſſed unnoticed when aſtronomy was leſs improved than at preſent, and it is alſo certain that many may deſcend to the ſun, and return, without a poſſibility of being ſeen by us; from which we may reaſonably conclude that their number is much greater than was thought in former ages. Theſe no doubt, have all their peculiar uſes, and are ſubſervient to many excellent purpoſes, tho' unknown to us, yet conſidered only as part of our ſolar ſyſtem, add not a little to its variety, beauty, and grandeur.

Canterbury Jan. 23, 1771.

JAMES SIX.

GENTLEMAN'S MAGAZINE *51* (1781), p.511.

A comet now visible described.

MR. URBAN, *Canterbury, Nov.* 24, 1781.

ON the evening of the 8th inſt. accidentally taking a view of the heavens, I ſaw ſomething like a nebulous ſtar near Urſa Minor, which, on viewing it with a teleſcope, appeared to be a comet. As ſoon as I had taken its diſtance from ſeveral ſtars, and marked its place on the globe, the ſky became cloudy, and I ſaw it no more that evening. This happened to be the caſe the three following nights, but on the 12th inſt. I ſaw it again, near the northern wing of Cygnus. The comet had moved above thirty degrees in four days, and paſſed very near the pole of the ecliptic into the ſign Aquarius. It has ſince croſſed the neck of Cygnus, the neck and left four foot of Vulpecula, and, on the 22d inſt. at five in the evening, it paſſed within 10′ of the ſtar ρ in the tip of the northern wing of Aquila. By the following particulars, any of your readers may deſcribe its courſe on the globe, and ſee the gradual decreaſe of its apparent motion:

	h		R. Aſc.		N. Decl.
1781, Nov. 8	— 10	——	254 — 20′	——	73° — 0′
12	— 11	——	290 — 40	——	44 — 30
15	— 5	——	295 — 20	——	31 — 20
18	— 6	——	298 — b	——	22 — 30
22	— 5	——	300 — 20	——	14 — 45

By the velocity of this ſmall comet's apparent motion, it ſeems to have paſſed not very far from the earth, and it is remarkable for having its courſe ſo nearly at right angles with the Ecliptic. In making an arc of 55° from the Pole of the Ecliptic, it varied its longitude very little. On ſome nights, the 22d inſt. in particular, I perceived a faint appearance of a tail, its coma extending about 15 in a direction oppoſite to the ſun. It is likely to be viſible ſome time longer, appearing to the naked eye like a ſmall nebulous ſtar. Though this comet reſembles, in ſome particulars, that of the year 1770, (which I had the pleaſure to ſee both approaching and returning from the ſun, and of which I gave an account in your volume for that year, pp. 319, 352); yet, as it croſſes the northern hemiſphere in a contrary direction, it is certainly not the ſame

JAMES SIX.

P.S. Any farther obſervations that I may make I will communicate to you next month.

GENTLEMAN'S MAGAZINE *53* (2) (1783), pp.726–7.

Of the great Heat

MR. URBAN, *Canterbury, Aug.* 4.

BY an extract of a letter from John Huxham, M. D. F. R. S. to William Watson, M. D. F. R. S. dated at Plymouth, July 19, 1757, it appears that 88 degrees, to which Fahrenheit's thermometer rose in the shade on the 12th of that month, was esteemed a very uncommon degree of heat of the air in England, which renders the unusual degree of heat we have lately experienced the more remarkable.

During the whole month of July the weather was extremely hot; on the 2d, as well as on the 28th of that month, thermometers in the shade, and properly exposed in the air, rose to above 87 degrees on Fahrenheit's scale; but on Saturday the 2d of this month, from 11 to 1 o'clock they stood at 90, and from 11 to 12 o'clock, as high as 92 degrees.

It is worthy of notice, that the very hot weather in 1757 was supposed to be the occasion of "much sickness, a "violent hurricane, and a smart shock "of an earthquake, at Falmouth, "Penryn, Truro, and thereabouts;" whereas I do not hear, that the late hot weather has produced any ill effects in the environs of this city, but gentle showers of rain have occasioned it to become much cooler, without any storms or violent commotions which have happened in many other places.

J. S.

*** Our correspondent's observation does not seem to agree with the generally received opinion; for it is believed that no year in man's memory ever abounded more with tempestuous weather, hurricanes, &c. than the present.

GENTLEMAN'S MAGAZINE *54* (1) (1784), pp.86–7.

Situation of the Georgium Sidus.

MR. URBAN, *Canterbury, Jan.* 26.

FOR the information of your astronomical readers, you receive the following particulars relating to the newly discovered (or Georgian) Planet.

Its longitude on the 24th inst. was $3^s\ 8^\circ\ 40'$; its latitude about 20′ N. It was in opposition on the 30th of Dec. last, and will be stationary on the 14th of March next. Its apparent daily motion is at present 2′ 30″ retrograde, gradually decreasing. It may at this time be seen every clear evening, without a glass, on the under part of Castor's right thigh, in the constellation Gemini, but cannot be viewed distinctly without a magnifying power of 60 or 70 times; and even then it appears in all respects like a star of the 5th or 6th magnitude.

J. S.

CAMBRIDGE UNIVERSITY LIBRARY MS COLL. ADD 7886/250.

JAMES SIX TO WILLIAM FRIEND *Canterbury, Jan. 29 1785*

DEAR SIR,

WITH this you will receive three Thermometer Indexes pack'd in glass tubes nearly full of spirit of wine and hermetically sealed, which was the easiest method I could think of sending them to you. Luckily I have by me the dimensions of the small tube of your Thermometer otherwise I should have been much at a loss.

The Indexes I send are fitted by a tube the same size as yours and are numbered 1, 2, 3, their springs differing accordingly. No.1 has the least pressure, and will barely support the index from falling down on the mercury. No.2 has a little more pressure, sufficient to answer every purpose. No.3 has still more and though it may be moved by two or three applications of a magnet of moderate power, yet on account of the stiffness of the spring there is some danger of the mercury passing by the Index and spoiling the Thermometer.

I would recommend No.2 to be first tri'd, as I apprehend it will best suit your purpose. If the spring should be too strong, No.1 is weaker if too weak No.3 is stronger but as the Indexes are of a very tender and delicate nature the following directions respecting the shifting them may not be improper.

The Caps being first taken off hang the Thermometer in a place of safety, in a good light, the top not above breast high and while the right hand is employed in taking out the Cork, the left may hold the cup out of which it is taken; least [*sic*] by too much force being used on that occasion the cup should be broken off from the smaller tubes on which it is set.

Taking now the tube containing the Index of the number you chuse; and with the magnet moving the Index (if it is not already there) to the end on which the cap and number is fixed, break the tube asunder near the other end by bending it short, opposite to a mark already made there with a file for the purpose; the Index may then be drawn out by the magnet. Holding now the magnet in the left hand carefully take hold of the Index with the thumb and forefinger of the right, but in so doing be very cautious not to let the finger touch the spring, but the lower Bugle and body of the Index only, for the glass spring being not larger than a hair is broken by the slightest touch: drop now the Index into the cup of the Thermometer and then by means of the magnet draw down into the smaller tube. If you wish to remove the Index from the magnet without putting it into the Thermometer, lay it on a piece of white paper, but do not attempt to take it up again but with a magnet.

I send you likewise two magnets, No.1 with rather less power than that which you have, and No.2 with rather more. Let the new Index be at first drawn down into the small tube with the weakest magnet No.1, because if on tryal you find the spring too stiff to be drawn down into the small tube you will probably want a greater magnetic power to draw it back again; keep therefore in reserve the stronger magnet for that purpose.

I flatter my self No.2 will suit you extremely well—let me know how you succeed—I wish you good luck—Adieu.

Dear Sir,
Yours in Sincerity
James Six

If you keep your magnets in the same place, lay their broadsides together parallel to each other, north to north and S to S poles. If they lay foot to foot, let the n. pole of one meet the s. pole of the other. But I would rather advise to keep them all in separate places least by meeting in a wrong position one should spoil the other. NB Be careful in unpacking the indexes which are in paper in one end of the roll.

TRANSACTIONS, AMERICAN PHIL. SOC., Vol.2 (1786).

N° XXXIV.

Astronomical Observations, communicated by Mr. RITTENHOUSE.

By Mr. *James Six*, of Canterbury.

Geocentric place of the New Planet.

April 1st, 1782,	29°	5′	30″	in ♊ lat. 13′ N.
October 15th,	7	21	18	♋ nearly stationary.
December 26th,	5	2	30	in opposition.
March 10th, 1783,	3	15	0	stationary.
October 15th,	11	53	10	stationary.
December 30th,	9	47	25	in opposition.
March 14th, 1784,	7	46	0	stationary.
				Lat. 23 N.

GENTLEMAN'S MAGAZINE 57 (1) (1787), p.18.

Six's *new Thermometer*.

MR. URBAN, *Jan.* 9.

IN the Philosophical Transactions for 1782 is a description of a new thermometer, invented by James Six, esq. of Canterbury. The distinguishing excellence and novelty of this instrument consists in its valuable quality of marking the greatest degree of heat and cold, *during the absence of the observer*; a quality which has ever been a grand *desideratum* in thermometers; though the means of producing this desirable effect, notwithstanding many attempts to accomplish it, remained unknown till the present very ingenious inventor, after much pains and attention, discovered them.

So useful an invention soon attracted the notice of foreigners. By the desire of a learned French professor, one of these thermometers was sent to Paris, and another to Italy; the latter of which coming to the sight of Signor Antonio Matteucci, an ingenious gentleman of Sienna in Tuscany, gave him so much satisfaction, that he has himself constructed several on the same plan; with the addition only of Reaumur's scale to that of Fahrenheit, for the convenience of his countrymen, who in general make use of the former. One of them he has had the honour of presenting to the Grand Duke of Tuscany, who, being himself both a lover of science, and an experimentalist, expressed great delight at the ingenuity of the invention.

As this anecdote will, I doubt not, give pleasure to all friends to the diffusion of useful knowledge, so, I hope, it may stimulate some of our mathematical instrument-makers (who are the best in the world) to make thermometers on this principle, that meteorological observers in general may be enabled to benefit by so useful an invention,

I cannot omit to add, that, in consequence of this, and other ingenious performances, publiſhed at different times, Mr. Six, ſome time ago, received, very unexpectediy, from America, a certificate of his inrollment as member of the American Philoſophical Society, eſtabliſhed at Philadelphia, of which Dr. Franklin is preſident.

Yours, &c. B. B.

GENTLEMAN'S MAGAZINE *59* (I) (1789), p.100.

Accurate observations of the mercury during the late frost.

Mr. URBAN, *Feb.* 13.

THE following accurate obſervations of the greateſt and leaſt heights of the mercury, during the late ſevere weather, may be acceptable to your readers. They were taken reſpectively from two Sixian thermometers, hanging without doors, one at Canterbury, and the other at Sienna in Italy, in lat. 43°. 10′.

Yours, &c. B. B.

	ENGLAND.		ITALY.	
1788.	Leaſt height in the night.	Greateſt height in the day.	Leaſt height in the night.	Greateſt height in the day.
Dec. 21	34	37	27	36
22	31	36	27	35
23	15½	28½	22	35
24	20	45	15	31
25	35	44	22	37
26	32½	44	31	30
27	21½	32½	29	33
28	21½	30	17	26
29	17	27	21	25
30	4½	22	15	23
31	4½	28	10	23
1789.				
Jan. 1	22	38	14	28
2	21	30	19	35
3	19	35	16	33
4	18	34	25	31
5	5½	27	21	34
6	10	28	19	31
7	21	26	19	33
8	71	28	23	35
9	11	25	38	43
10	15½	32½	38	45
11	25	34	44	53

GENTLEMAN'S MAGAZINE *63* (2) (1793), p.864.

Obituary

At his houſe in Canterbury, James Six, eſq. F.R.S. a gentleman well known to the lovers of ſcience for his ingenious and ſucceſsful purſuits in aſtronomy and natural philoſophy; and whoſe natural genius and various abilities diſtinguiſhed him, from the *many*, among thoſe who were intimate with him, though his modeſty kept him from that diſtinction his merit deſerved. He was one of the French refugee families who ſettled at Canterbury in the reign of Queen Elizabeth, on account of the perſecution of the Proteſtants in France, and eſtabliſhed the ſilk-trade there. Mr. S. was brought up to that buſineſs, but, on its decline, retired, early in life, on a handſome competency, to purſue his love of ſcience, and with care to educate his ſon and daughter. He was an eminent aſtronomer, having frequently made very accurate diſcoveries in that ſublime ſtudy, which he communicated by occaſionally correſponding with other aſtronomers all over Europe. He preſented to the Royal Society an improved thermometer of his invention, deſcribed in their Tranſactions, vol. LXXII.; and an account of ſome experiments to inveſtigate the variation of local heat, inſerted in vol. LXXIV. of the ſame work. Theſe communications procured him the honour of admiſſion into that learned body. He alſo made ſome uſeful experiments in electricity; and, having very good and expenſive electrical machines, which he perfectly underſtood, he was ever ready to aſſiſt with his practical ſkill whenever the medical opinion deemed that electrifying could be beneficial. He was alſo well known as a floriſt, and could, with peculiar beauty, taſte, and preciſion, paint the flowers he had reared, as well as uſe the pencil elegantly on other ſubjects, for amuſement. His mind thus replete with variety of knowledge, he filled up the meaſure of his time, without having much to beſtow, on general ſociety, though his diſpoſition was chearful, communicative, and philanthropic in the higheſt degree, which was evinced by his indefatigable attention to a Sunday-ſchool, *principally* inſtituted, and zealouſly patroniſed, by him; which more eſpecially he laboured to improve, and to which he dedicated his leiſure hours more ſteadily, as a pious work to engage his thoughts at a period when the ſevere ſtroke of loſing his only and very valuable and accompliſhed ſon oppreſſed his heart with the deepeſt ſorrow.—James Six, jun. M.A. of Trinity-college, Cambridge, had travelled with the ſon of Sir J. Stanley, and was juſtly and highly eſteemed by that family. He travelled a ſecond time, and died, early in life, of a fever, at Rome, Dec. 14 1786; where he was buried with peculiar honours, notwithſtanding the general ſtrictneſs of the Romiſh Church (ſee our vol LVI. pp. 72, 90). His character being univerſally reſpected, a monument was erected there by Sir John Stanley, unknown to the parents of young Mr. Six, who have placed an elegant one in Weſtgate church, Canterbury; to which alſo Mr. Six, with another gentleman, have given an organ, which accompanied the voices of the little choriſters, who were foſtered by his care, encouraged in piety and induſtry by his precepts, and indulged by his benevolence in whatever contributed to their well-doing in their humble ſtation. By them and many more he will long be ſincerely lamented, for he was a good man, an uſeful member of ſociety, and a good Chriſtian. He has left a diſconſolate widow, and an only daughter, truly worthy of ſuch a father's fond affection, and married to Mr. May, an opulent brewer at Maidſtone.

BIBLIOGRAPHY OF JAMES SIX'S PRINTED WORKS

Observations on the appearance of the late comet. *Gent. Mag. 40* (1770), pp.252–3.
Observations on the re-appearance of the late comet. *Gent. Mag. 40* (1770), pp.319–20.
Six's account of a late comet. *Gent. Mag. 41* (1771), pp.28–9.
A comet now visible described. *Gent. Mag. 51* (1781), p.511.
An account of an improved thermometer. *Phil. Trans. 72* (1782), pp.72–81.
Of the great heat. *Gent. Mag. 53* (2) (1783), pp.726–7.
Situation of the Georgium Sidus. *Gent. Mag. 54* (1) (1784), pp.86–7.
Experiments to investigate the variation of local heat. *Phil. Trans. 74* (1784), pp.428–36.
Astronomical observations . . . by Mr James Six of Canterbury. *Trans. Amer. Phil. Soc. 2* (1786), p.260.
Experiments on local heat. *Phil. Trans. 78* (1788), pp.103–20.
The construction and use of a thermometer. Maidstone (1794).

THE CONSTRUCTION AND USE OF A THERMOMETER.

WITH EXPERIMENTS, &c.

THE CONSTRUCTION AND USE OF A THERMOMETER,

FOR SHEWING THE EXTREMES OF

TEMPERATURE IN THE ATMOSPHERE,

During the Observer's Absence.

TOGETHER WITH

EXPERIMENTS ON THE VARIATIONS

OF

LOCAL HEAT;

AND OTHER

METEOROLOGICAL OBSERVATIONS.

BY JAMES SIX, ESQ. F.R.S.

MAIDSTONE:
Printed and Sold by J. BLAKE; and Messrs. G. & T. WILKIE, LONDON.
1794.

PREFACE.

THE disadvantages to which posthumous publications are liable, give them a peculiar claim to indulgence. As the hand that sketched them can no more be raised to supply their deficiencies, or to retrench their excess, their merit, however shaded by imperfection, will plead in mitigation of critical severity. When a composition is left not complete for the press, it must want many little touches and additions, which the taste or skill of the original writer would have suggested for its improvement. And it can rarely happen that the person, to whose care the printing of a work devolves, is so versed in the subject it discusses, and so devoted to it, as to be equally competent with the author to its correction.

This

This remark will apply with great propriety to the following publication. For though no reaſonable doubt can be entertained of its favourable reception, as a valuable acquiſition to the ſcience of Meteorology, which can be advanced only by the aid of well authenticated data; yet it is certain, that had Mr. Six lived to ſuperintend its printing, it would have received many additions, and have been a more accurate performance than the publiſher's inexperience allows him to hope it will now be found. That it is leſs perfect than it might have been, was not deemed a ſufficient cauſe for withholding it from the public. That it appears ſo correct as it is, muſt be attributed to the reviſal of a Reverend Friend of the Author, whoſe zeal and ſucceſs in the promotion of philoſophical enquiries, the learned world has long ſince acknowledged and approved.

With reſpect to the writer of the papers here collected, it may juſtly be affirmed that he was a man of ſtrong natural abilities, which he cultivated with great care, and

and diſtinguiſhed himſelf through life by his integrity, ſimplicity, and candour. Endued with a capacity for exploring the depths of ſcience, he had an underſtanding above the pride and parade of it. Though naturally unobtruſive and retired, he was not unknown as a philoſopher; and his correſpondence on ſcientific ſubjects procured him the honour of being elected a member of our Royal Society, and of that eſtabliſhed at Philadelphia. He watched and well knew the ſtate and progreſs of modern diſcoveries in every branch of knowledge; and his talents were particularly and uſefully employed in the ſtudy of Aſtronomy, the Mathematics, and Natural Philoſophy in general. Of Painting, as a ſcience, few men underſtood more correctly the principles and theory; and his own performances afford no mean proof of his elegant taſte, and proficiency in the art.

His memory will long be dear to thoſe who aſſociated with him in his private hour, the witneſſes of his moral and religious conduct. It is this part of his character

that

that conſtitutes his higheſt praiſe; a character in which unaffected piety, benevolence without oſtentation, and an engaging ſuavity of manners were eminently conſpicuous. So far was he from the ſcepticiſm, which the courteſy of the age is too apt to ſuppoſe the reſult of deep thinking, that he was earneſt in exerting the influence, which ſuperior talents always command, in maintaining the neceſſity, truth and certainty of revelation. In whatever he was employed, his object was to be ſerviceable. From the moſt abſtruſe ſpeculations he could turn with ſatisfaction to the ſocial duties of life, to the diſcharge of thoſe numberleſs good offices, which neighbourhood invites and humanity demands. In ſuch exerciſes his activity was inſuperable: and his ſolicitous patronage of the Sunday School, founded in his pariſh in Canterbury, to the ſuperintendance of which he devoted a very conſiderable portion of his time and attention, will be a laſting monument of his piety, beneficence, and patriotiſm.

Happy

Happy it was that his mind was open to the consolations of religion, when afflicted with the loss of an only son, whose virtues and accomplishments answered his fondest wishes; who in his twenty-ninth year died, and was interred in Italy. It need not therefore be told with what exemplary fortitude and resignation he supported that most severe infliction; nor can the merit of his amiable son derive any additional lustre from the attempt of a feeble, and it may be said, a partial encomiast. Suffice it to refer to a slight account of him given in the * Obituary of the Gentleman's Mazine

* "At Rome, JAMES SIX, M. A. Fellow of Trinity College, Cambridge. He was a young man of great natural abilities, and of extensive learning. He understood the Hebrew, Greek, Latin, Italian, French, and German languages, and in most, if not all of them, had a well-grounded and accurate knowledge: of his classical and mathematical learning, the several prizes which he obtained during the course of his academical studies, are an eminent and honourable proof. Two beautiful odes (vol. LIV. p 285, 286,) translated from the German, give no mean idea of his poetical powers; and, as a draughtsman, his designs were executed with wonderful neatness and elegance. To these accomplishments, which adorn society, he added a sweetness of manners, and a benevolence of disposition, that endeared him to his family and friends, and gained him, wheresoever he went, attention and esteem. He was buried at Rome [see p. 72], in a place appropriated

gazine for January, 1787, which notices in reſpectful terms both theſe truly excellent characters.

" Animæ, quales neque candidiores
Terra tulit, neque queis me ſit devinctior
alter."

Should the following ſheets prove an acceptable preſent to the public, it will pardon the preſumption of the Editor in laying before it theſe ſhort and imperfect memorials

priated to the burial of Proteſtants, and the funeral ſervice was read over him by the Rev. Mr. Waleſby, chaplain to his Royal Highneſs the Duke of Glouceſter: the mournful ceremony was attended by Sir Cecil Biſhop, Sir Thomas Stapleton, Mr. Long, and moſt of the Engliſh gentlemen then reſident at Rome. He was the ſon of Mr. JAMES SIX, of Canterbury, to whoſe ingenious obſervations and experiments in natural philoſophy, &c. the public have been much indebted." [*Gentleman's Mag. for January*, 1787.]

As a taſte for German literature ſeems to be advancing in this country, it may not be improper to add to this account, that, a ſhort time before his death, he completed a tranſlation of the Oberon of Mr. WIELAND, a poetical romance, in 14 Books. But ſome doubts of the reception which the tranſlation of a German poem might meet with in England, which were increaſed by the opinion of Mr. WIELAND himſelf (who nevertheleſs expreſſed in handſome terms his approbation of ſome parts of it which he had ſeen) as they unfixed his reſolution to print it, have likewiſe ſince his death prevented its publication.

rials of an eminently worthy person, and indulge him in the melancholy satisfaction of thus exhibiting his regard to the memory of the Author, by whose death society has lost a valuable member, and himself a Friend, a Father.*

Maidstone, May, 1,
1794.

* Mr. SIX died August 25, 1793, aged 62.

CONTENTS.

CHAP.

INTRODUCTION.

THERE is not, perhaps, any philosophical instrument more generally in use than the Thermometer. The construction and improvement of it has engaged the attention of Men of the most distinguished abilities.

To render it, however, completely fit for Meteorological observations, to which, from its first invention, it has been constantly applied, an essential property was still wanting, namely, to shew the greatest and least degrees of Heat in the absence of the observer.

The inconvenience, as well as inaccuracy, attending the use of the common mercurial Thermometer, is very obvious. For as the height of the mercury can only be known while the eye is on the instrument; and as the precise time when the extremes either of Heat or Cold happen, is uncertain, it is evident that frequent errors in the observations will be unavoidable.

To remedy this defect several attempts have been made. In 1689, the celebrated JOHN BERNOULLI proposed to Mr. LEIBNITZ the plan of a Thermometer* to shew the greatest and least heights of the mercury in the observer's absence. Mr. KRAFT in 1740, without knowing of Mr. BERNOULLI's invention, constructed one nearly like it, which he also adapted to the purpose of shewing the temperature of the Sea. In 1757 and 1758, Lord CHARLES CAVENDISH presented the Royal Society with designs of a Thermometer for the same purposes, but on a different plan.† Mr. KEAN FITZGERALD also in 1760, invented one to shew the extreme variation of temperature of the Air in the observer's absence.§

The ingenuity which appears in the contrivance of these several instruments may be seen in the descriptions that have been published; and

* Diss. sur la Comparaison des Therm. par VAN SWINDEN, p. 253—255.

† Phil. Trans. vol. L. p. 501.

§ Phil. Trans. vol. LI. p. 820.

to which the author begs to refer his reader, rather than to point out what may seem to him defective in their construction.

Apprehending, however, that a Thermometer might be contrived more conveniently and with greater accuracy to accomplish the purposes before mentioned, he applied himself to construct one on a principle different from any of the former. And although his account of this instrument (as well as another to shew the temperature of the Sea) has been already published in the Philosophical Transactions;* yet, since most of those made in imitation of the former are essentially defective in their construction, and the author's account of the latter not so complete as he is now able to give, he has judged it necessary to publish this still more particular description of them both, with some experiments and remarks on the general method of making Meteorological observations with Thermometers.

In the Appendix the reader will find such information respecting the manner of constructing these Thermometers, and also of repairing any damage they may accidentally receive, as may be useful either to the Artist or Philosopher.

* Vol. LXXII. p. 80.

The CONSTRUCTION, &c.

CHAP. I.

DESCRIPTION OF A THERMOMETER TO SHEW THE EXTREME VARIATIONS OF TEMPERATURE IN THE OBSERVER'S ABSENCE.

PLATE 1. Fig. 1. *a. b.* is a tube of thin glass 16 inches long, and a quarter of an inch in diameter: *c. d. e. f. g. h.* a smaller tube, with the inner diameter $\frac{1}{16}$ joined to the larger at the upper end *b.* and after rising about an inch above it, bent downwards on one side, and from two inches below the end *a.* upwards again on the other side, in the direction *c. d. e. f. g. h. i.* parallel to, and about half an inch distant from it, but not perfectly in the same plane. At the end of the small tube from *h.* to *i.* two inches and a half in length, the inner diameter is enlarged to $\frac{3}{8}$, and the end is closed by a brass cap. This glass is filled with highly rectified spirit of wine, to within an inch

 and

and a half of the end, excepting that part of the small tube from *d.* to *g.* which is filled with mercury.

From a view of the instrument in this state it will evidently appear, that when the spirit in the large tube, which serves as the body of the Thermometer, is expanded by Heat, the surface of the mercury at *d.* will be pressed down, and consequently the surface at *g.* will rise; and when it is condensed by Cold the reverse will happen: so that whenever the spirit is either contracted or dilated by the change of temperature, the mercury, on one side or the other, will always rise. The scale, which is *FARHENHEIT's, beginning with o, on the top of that side where the mercury rises by increase of Cold, has the degrees numbered downwards; while that on the opposite side, where the mercury rises by increase of heat, has the degrees numbered upwards.

The divisions of the scale are ascertained by placing the Thermometer, with a standard mercurial one, in a large vessel of water, heated or

* Because this scale is generally used in England. Signor Antonio Matteucci, who has made fome of thefe Thermometers in Italy, from one of mine, which I sent to Mr. Grinfield of Sienna in Tuscany, added to the scale of Farhenheit, that of Reaumur, which latter is mostly in use on the Continent.

cooled

cooled to whatever degree may be found necessary, and marking them at every five or ten degrees.

Within the small tube of the Thermometer, above the surface of the mercury, immersed in the spirit of wine, is placed, on either side, a small index, so fitted as to be moved up and down as occasion may require. The surface of the mercury when it rises, carries this index up with it, which not returning with the mercury when it descends, but remaining fixed by means of a spring, accurately shews how high the mercury had risen, and consequently the greatest degrees of heat and cold that have happened. Fig. 3. represents one of these indexes drawn larger than the real one, to render it more distinct. *a.* is a very small glass tube $\frac{5}{8}$ of an inch long, hermetically sealed at each end, inclosing a piece of fine steel wire, nearly of the same length. The ends *b. c.* are covered with small pieces of black glass or bugles, of such a diameter as just to pass freely up and down within the small tube of the Thermometer. From the end of the index at *b.* is drawn a spring of glass to the fineness of a hair, about an inch and a half in length, which being set a little obliquely, presses against the inner surface of the tube, and prevents the index from following the mercury when it descends, or being moved by the spirit passing up and down, or by any sudden motion given to the instrument by the hand or otherwise. But at the same time the pressure is so adjusted as to permit the index to be readily carried up by the surface of the rising mercury, and downward whenever the instrument is to be rectified for observation.

Fig. 2. represents the Thermometer on its frame. The plates, on which the scale is graven, are made to slide out; and the frame is open at the back, by which means the body of the Thermometer is in a great measure insulated. The cap *a*. and the base *b*. are made to fix on with screws, and only cover the end and turnings of the small tube. Small pieces are fixed to the back of the frame, to prevent the whole of it from coming in contact with the wall, or place against which it is suspended; and the lower end of the frame should be so confined when it hangs without doors, as to prevent its being shaken by violent winds.

When the Thermometer is to be used for observation a small magnet is to be applied to that part of the tube against which the indexes rest, by which they may be immediately brought down

Pl. II.

Fig. 1.

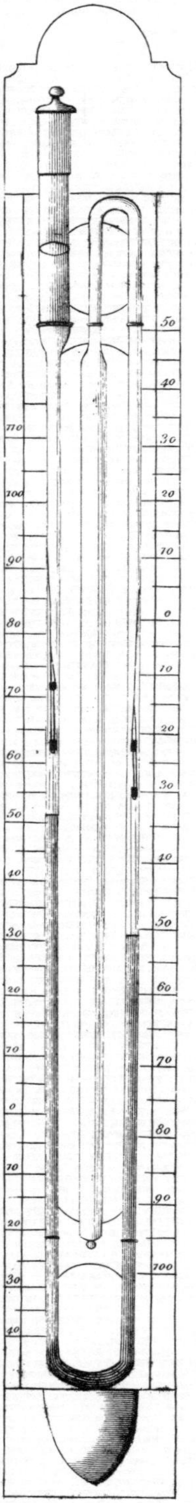

Fig. 2.

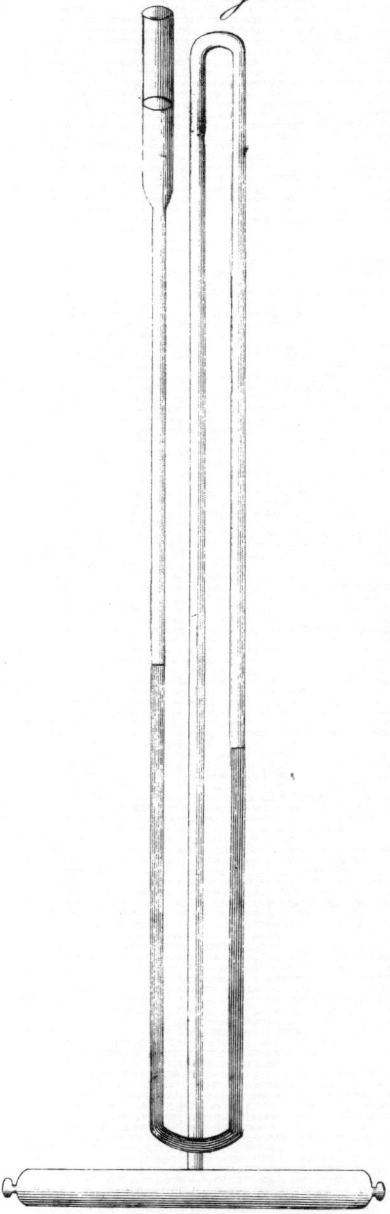

down to the surface of the mercury, and the instrument thereby rectified. The indexes, though of a tender and delicate nature, are not, when once placed in the tube, liable to be broken; and the thermometer may be exposed to all kinds of weather, without receiving the least essential injury.

Pl. 2. Fig. 1. is a Thermometer of the same form, but of smaller dimensions; the body of it being only ten inches in length, and $\frac{20}{90}$ in diameter; and the neck or smaller part $\frac{5}{90}$. At two inches and a half from the end the diameter is enlarged to a quarter of an inch. This small tube is carried round the body *in the same plane* at $\frac{2}{10}$ of an inch from it, and closed at the end with a brass cap. The frame of this smaller size is not convex in the middle, as the large ones; but perfectly flat, with the scale graven upon it, and so far pierced and open in the middle that the whole glass is in a great measure insulated: neither is it covered at the ends like the other, but left entirely open. This instrument, when used for observation, should be kept about two inches from the wall, or place, against which it is suspended.

Pl. 2.

Pl. 2. Fig. 2. is a Thermometer of the same kind, with this difference only, that the body is but six inches long, and of such a diameter as to contain the same quantity of spirit as fig. 1. and placed in a horizontal position, for the convenience of applying it very near the ground, or surface of any other bodies. But as Thermometer fig. 1. pl. 2. may be placed to within 10 or 20 degrees of the horizon, I have found it rather more convenient for most experiments on or near the ground, than this.*

These three instruments are essentially the same: the two first differing only in the size, and the form of their frames, the other in the horizontal position of the body.

* This Thermometer is drawn without a frame, that the disposition of the tubes may be more distinctly seen. The frame may be made as best suits the experiments for which it is intended.

CHAP. II.

REMARKS ON THE CONSTRUCTION OF THIS THERMOMETER.

NOW it evidently appears that this Thermometer requires a greater quantity of fluid than the common ones, to give sufficient motion to a column of mercury large enough to carry up indexes on its surface. The body is therefore filled with spirit of wine, that fluid expanding much more than mercury, and being also much lighter: and that the spirit may be more readily affected by change of temperature, the form given to the body is that of a slender tube. The smaller tube or neck of the Thermometer, instead of being carried upright, like the common ones, is bent down on one side of the body, and up again on the other, which not only renders the instrument compact, but is also in other respects essentially

sentially necessary: for by these means the column of mercury from *d.* to *g.* like an inverted syphon, has its two ends or surfaces turned upwards, on which the indexes alternately ascend. The curvature from *e.* to *f.* is also necessary for the column of mercury on each side to rest upon, and to prevent it from shifting its place in the spirits, which without it would certainly happen. For if too great a degree of heat should cause all the mercury on the side *c. d. e.* entirely to pass through the curvature *e. f.* into the side *f. g. h.* the whole column then in a vertical position, without any other support than the spirit, would gradually sink down to the curvature; and the spirit passing between the mercury and the glass, would rise above it: but as long as the mercury remains in the curvature *e. f.* though all the rest of the column should be on one side, and even remain so for a considerable time, the mercury would, nevertheless, retain its place in the spirit.

That part of the spirit in the small tube from *g.* to *h.* has no effect on the motion of the mercury; and is placed there only to diminish the friction of the index spring, immersed in it, and render it more equal to that on the other side: but care should be taken to leave a sufficient quantity of air at the end *i.* for whenever the spirit

rit in the body of the instrument is contracted by cold, the rising of the mercury, as well as the spirit, in the side *c, d. e.* depends on the pressure of the air at that end. The pressure of the external air would readily accomplish this; but if the tube was left open for that purpose, particles of dust would get into it, and obstruct the rising of the index, and the spirit would be lost by evaporation; the extremity *i.* therefore, is closed by means of a brass cap to open occasionally; and the elasticity of the inclosed air is found sufficient for the purpose.

Insulating the body, and keeping the frame from touching the place against which it hangs, is more necessary than at first sight may appear.

By means of the steel wire within the index, the instrument may be rectified in the most commodious manner, whenever the magnet is applied; nor is it liable to be corroded, being inclosed in glass, which on account of its elasticity is also extremely proper for the spring, and is not in danger of suffering the least alteration from remaining always immersed in the spirit.

The

The bugles, or small pieces of black glass at each end, being of a larger diameter than the body of the index, are necessary for keeping it parallel to the sides of the Thermometrical tube; and their dark colour and sharp edge shew very distinctly on the scale, the degree to which it has been elevated. The small knobs at each end, formed by sealing the glass hermetically, have likewise their particular use. That at the lower end not only keeps the bugle fast on the tube; but also by presenting itself first to the convex surface of the rising mercury, the index is made to float thereon sooner than it would otherwise do, and the mercury is thereby prevented from passing by it. The little knob at the extremity of the spring, by its spherical and smooth surface, contributes to diminish the friction, as well as render it more equal.

CHAP. III.

EXAMINATION OF SOME OBJECTIONS WHICH HAVE BEEN MADE TO THE CONSTRUCTION OF THIS THERMOMETER

AND first, because it contains mercury and spirit of wine*, which different fluids when heated, expand in a different ratio, some persons have concluded that the regularity and truth of the scale must be destroyed, and will not, in every part of it, agree with Thermometers made with mercury, or one fluid only. But this objection will appear groundless, if we attend to the method by which the scale is made. It has been already observed, that for this purpose, the instrument is wholly immersed in a large vessel of

* The quantity of mercury bears a very ſmall proportion to that of the spirit.

water, with a standard mercurial one, by which every fifth degree is taken. So that however differently the fluids or solid parts, of which the Thermometer is composed, may be affected by change of temperature, they all, according to their several kinds, expand and contract uniformly in the same degrees of heat, and in determining the scale mutually combine always to give the mercury the same motion.†

On presenting one of these Thermometers to the Royal Society, they did me the honour to publish my account of it in their Transactions. The Critical Reviewers conclude their remarks on that paper, by saying "the machine is too complica-"ted, and the resistance of the index, on one hand, "and the necessary bulk of the spirits on the "other, seem to us very material objections." I agree that simplicity is an excellence in the construction of any instrument; yet there are machines much more complicated than this Thermometer, which, nevertheless accomplish the purposes for which they are made, in the most commodious, accurate, and elegant manner.

† If this objection had any weight, it would apply more properly to the glass which contains the fluids, and which is common to all Thermometers; for the expansion of the glass by heat, directly counteracts the rising of the fluids.

That

That a single instrument which indicates the extremes of both heat and cold in the observer's absence, should be altogether of as simple a construction, as one that shews neither, is more than ought, in reason, to be expected; nor is it necessary that it should, provided the parts of which it is composed, be, with regard to matter and form, so adapted and combined, as completely to answer its design.

By *resistance of the index* is meant, probably, either that the rising of the mercury will be, in some measure, obstructed or retarded by it; or, that the index, by the pressure of its spring, or otherwise, will be prevented from being carried up on the surface of the rising mercury, so readily as it ought to be. But this can never happen; because the indexes, independent of the spring, are so fitted as to pass freely up and down in the spirit; consequently sufficient space can never be wanting for the mercury to pass by them, if they did not readily float on its surface. The pressure of the spring, in this case, acts as a small addition to the weight of the index, and is therefore adjusted accordingly; but will no more obstruct the rising of the mercury, than vessels, left on the sea-shore by the tide, will obstruct the rising of it when it returns. And even if it were possible for the motion of the fluid to be at all retarded by any resistance of the index, such resistance being uniform, no error could hence arise, the scale being made under the influence of every cause, which can operate, either to accelerate or retard its motion.

Lastly, *the necessary bulk of the spirits* seems to imply that some inconvenience must arise from the body of the Thermometer and quantity of spirits being too large; and since large compact bodies are not so soon affected by change of heat and cold as small ones, it is therefore concluded that this Thermometer will not so readily shew any change of temperature as it ought to do.

Now it is well known that the readiness with which bodies are liable to be affected by heat and cold, depends more on their relative surface, than their quantity. The quantity of spirit necessary for one of these largest Thermometers would something more than fill a spherical bulb of an inch and a quarter in diameter; which form would confessedly have too small a proportion of surface, and be too slowly affected by change of Temperature. To obviate this, I have extended the spirit in a very thin glass tube, sixteen inches long, and but little more than a quarter of an inch in diameter;

eter; whereby the surface is so considerably enlarged, as to be immediately sensible of the least variation of temperature in the air.

We may, therefore, I presume, fairly conclude—That the machine is not *too complicated*, since it commodiously and accurately accomplishes the purposes, for which it was constructed.

That, the indexes being properly made, will be readily carried up on the surface of the mercury, and cannot at all retard, or occasion the least *resistance* to the rising of it.

And, that any inconvenience which might arise from the *bulk of the spirits*, is obviated by the extension of their surface.

Consequently, That the objections, supposed to be very material, are founded in misapprehension only.

CHAP. IV.

THE MANNER OF USING THIS THERMOMETER; WITH REMARKS ON THE GENERAL METHOD OF MAKING OBSERVATIONS ON THE TEMPERATURE OF THE AIR WITH THERMOMETERS.

THE better to form a judgment in this matter, it may not be improper, briefly to consider the cause of heat, and the variations of it in the lower regions of the atmosphere.

The action of the Sun's rays on the surface of the earth, appears to be the principal cause of heat; because those parts which are most exposed to the Sun, experience in general the greatest degrees of it; and as the several parts of the earth, by its annual and diurnal motion, are differently exposed to the Sun, so is their temperature varied accordingly.

Tis

'Tis true the degree of heat at different places does not exactly correspond with their respective latitudes; nor the variation of it uniformly follow the astronomical motions of the earth. Variety of inferior causes contribute, in a considerable degree, to interrupt that regularity. A constant direction of winds from hotter regions; the vicinity of seas and oceans, extensive continents, high mountains, land more or less cultivated, or covered with wood; these causes affect the general temperature of the places where they are found. The change of winds, or an upper and lower current blowing in different directions; the clearness or cloudiness of the sky; rain; evaporation from the surface of the earth; the ascent, descent, rarefaction, and condensation of vapours; the no less powerful, though imperceptible, operations of electricity; all these, severally or in combination with each other, frequently produce great and sudden changes of temperature in the air, and prevent that regularity which the annual and diurnal revolutions of the earth would produce. In these cases, however, it is often difficult to distinguish the cause from the effect. Upon the whole it is evident that a general change of temperature, however uncertain with respect to the precise time and exact degree, must certainly take place within the course of twenty-four hours; consequently the principal object in keeping a thermometrical journal, is to note the greatest and least degrees of heat which happen within that period.

The general method of keeping these journals is seen by those which are daily published. They give for every day the greatest and least heights of the mercury taken at stated hours; from the aggregate of which, either a monthly or annual mean is usually taken, which is called the monthly or annual mean temperature of the place. But from the uncertain combination and irregular effects produced by the causes before mentioned, the extreme degrees of heat and cold in the twenty four hours, do not happen always exactly at the same time, especially the greatest degree of cold, which is generally taken too late in the day. Such observations, therefore, by not giving the true extremes, cannot give the true mean.

That journals are generally formed from observations made in this manner, is certainly not matter of choice, but necessity; for the observer can never know the height of the mercury in the common thermometer, except when his eye is upon it; and as that cannot be for any considerable time, it is necessary for him to confine his obser-

vations

vations to stated hours, and he must remain totally unacquainted with any greater degree of heat or cold which may have happened during the night, or in his absence.

To find the errors which will arise from calculating the mean temperature of a place, from observations taken at the hours which have been usually chosen for that purpose, I made a course of observations of nearly the whole month of May at six and seven o'clock in the morning; and in September and October at seven and eight in the morning, and at ten in the evening. A mean taken from the sum of all the heights, I found to differ about three degrees from the mean taken from observations made at the same time, and with the same instrument, the indexes of which gave the true maximum and minimum heights of the mercury. A difference which will certainly happen, more or less, when the least height of the mercury is not taken into the account; and which it cannot be by the common method.

The general taste for natural science which universally prevails, has induced the curious, in different parts of the world, to keep meteorological journals; the first and most obvious use of which is, by comparing them together, to endea vour to form some general theory of heat and cold, with the variations of it, in every latitude, and at every season of the year.

Mr. KIRWAN and the celebrated PERE COTTE have published to this purpose a great number of observations, collected from the memoirs of different academies. The first of these ingenious writers has calculated tables, as well of the monthly as of the annual mean temperature of every latitude, from 10° to 80°. Not that he gives it as a perfect calculation, to be in all respects depended on; but only as an attempt to reduce, as near as possible, to regular theory, the general heat and cold of different places and latitudes. By these tables, the annual mean temperature of lat. 51° is 52.4; and by observations, which he has given, from the year 1772 to 1780, made at the house of the Royal Society in London, lat. 51° 32′ the mean temperature of London is 51.9. or in round numbers 52°, nearly agreeing with his table. But this, I presume, is stated too high; for from the year 1780, to 1791, by observations that can fully be depended on, the mean temperature has been found to be only 50.61, which is the result of observations taken at stated hours. And I have reason to believe that if the true minimum nocturnal height of the mercury

cury were taken into the account, the mean temperature of London and its environs would be found to be nearly three degrees lower. I form my conjecture from the temperature of Canterbury; which, from its nearer vicinity to the sea, with which, at the distance of from six to sixteen miles it is nearly surrounded, should be something warmer than London. But from a series of eleven years observations with my own Thermometer, I find the mean temperature 47. 9. Had I taken my observations at stated hours, as those were taken in London, it is evident from the experiments which I have before mentioned, that my mean temperature would have been about three degrees higher, or 50. 90. nearly what was found to be the temperature at London.

CHAP.

CHAP. V.

HOW FAR A THERMOMETER IS LIABLE TO BE AFFECTED, BY BEING PLACED NEAR TO, OR AT A DISTANCE FROM THE EARTH; OR, WHAT DIFFERENCE IS FOUND IN THE TEMPERATURE OF THE AIR AT DIFFERENT ALTITUDES.

THE information I obtained in this matter, was by the following means. I took corresponding observations at different stations, from the surface of the earth, to an elevation of more than two hundred feet above it. On making this experiment, I was at first very much surprised to find the temperature at the upper station, on some nights considerably warmer than the lower one; and on searching for the cause of this difference, I found it to proceed from a refrigeration, which takes place on the surface of the ground in the evening and night, and more particularly so, when the weather

weather is still and clear. For although the earth is then more liable to be heated by the sun's rays, in the day time, and by its superior density capable of retaining a greater degree of it than the air; yet a diminution of heat begins to take place on its surface, as soon as the sun is near setting, and the dew begins to appear; and increases so as considerably to cool the lower strata of the air, according as they are nearer to, or farther distant from it; and this nocturnal diminution of heat is always greatest in severe frosty weather.

To see whether this Phenomenon varied with the different seasons of the year, I continued for above a twelvemonth, to place Thermometers at different altitudes,—on Canterbury Cathedral, and others nearer to the ground. The most considerable differences I found during that period, as well as at other particular times, when the weather was serene, may be seen by the following table.

The Thermometers I made use of on this occasion were such as I have already described, and when placed together agreed with each other; but for the purpose of making this experiment, one of them was suspended on the Cathedral tower, 220 feet, and the other in my garden, 9 feet only, from the ground; both in an open shady northern aspect. The nocturnal degrees of cold, as stated in the table, are those of the night immediately preceding the day to which they are affixed.

	In the Garden. at 7 feet.		On the Tower, at 220 feet.		Difference	
	Night	Day	Night	Day	of Night.	of Day
1784						
Jan. 10	9	21	15	20	6	1
11	5	22	12	25	7	+ 3
1785						
May 9	36	70	45	68	9	2
10	36	62	44	61	8	1
June 13	41	67	49	64	8	3
14	$42\frac{1}{2}$	69	$50\frac{1}{2}$	64	8	5
25	40	62	48	58	8	4
27	43	67	52	62	9	5
July 13	46	67	52	64	6	3
Oct. 5	41	$57\frac{1}{2}$	51	56	10	$1\frac{1}{2}$
6	41	56	47	55	6	1
7	40	$54\frac{1}{2}$	46	52	6	$2\frac{1}{2}$
8	40	55	$46\frac{1}{2}$	53	$6\frac{1}{2}$	2
16	$28\frac{1}{2}$	53	$38\frac{1}{2}$	51	10	2
17	28	54	$38\frac{1}{2}$	$52\frac{1}{2}$	$10\frac{1}{2}$	$1\frac{1}{2}$
18	29	56	$38\frac{1}{2}$	53	$9\frac{1}{2}$	3
1786						
Jan. 4	$-2\frac{1}{2}$	30	15	31	$17\frac{1}{2}$	+ 1
Nov. 27	39	47	44	47	5	0
Dec. 12	24	38	31	37	7	1
26	6	31	17	32	11	+ 1
27	$6\frac{1}{2}$	44	21	42	$14\frac{1}{2}$	2
29	27	45	34	46	7	+ 1
1787						
Feb. 6	31	43	36	43	5	0

notwith-

Notwithstanding the excess of cold which appears to have happened on the *particular nights*, inserted in this account, when the weather was perfectly still and serene; yet at other times I have found little difference between the two stations; and that little in summer, at the upper one in the day time, inclined towards excess of heat; so that the monthly mean temperature of the two stations never differed so much as two degrees, and the annual mean only 0. 44; the upper station being on the whole not half a degree colder than the lower one. It is however worthy notice, that from the beginning of March to the end of August, the monthly mean of the Tower was colder than the garden, and the other six months it was warmer, so far was the variation in general affected by the different seasons of summer and winter.

And here I shall beg leave to take notice of a work published at Geneva in 1790, wherein the author M. A. Pictet, has done me the honour to mention my experiments. After speaking of his own on the temperature of the air at different heights, he says "Elles ont èté repetèes dès-lors pars un Physicien Anglois M. Six, et les resultats sont conformes à ceux qu j'avois obtenus." for although, as Mr. Pictet truly observes, an account of his experiments had been published some years before

before in a work of Mr. de Luc's, yet I must confess I remained wholly unacquainted with them till after most of mine were made.

In respect to the similarity in the result of our observations, it appears by his account that we were, in the first instance, both equally surprised to find the temperature at the upper station, frequently warmer than the lower one; and upon examination observed this variation to proceed from a diminution of heat, which takes place on the surface of the earth, in the night time, under the same circumstances and disposition of the atmosphere, at Geneva, as in England; and that the range of the Thermometer in clear weather was greater at the lower station than at the upper one. Yet in other particulars, as well in the result, as in the manner of making the experiments, we seem to differ.

Mr. Pictet's observations were made with mercurial Thermometers, placed at 5, at 50, and afterwards at 75 feet from the ground. The upper one was supported in its station by a mast or pole, of sufficient length, erected on a plain, and braced to the ground with cords; to the top of this mast the Thermometer was drawn up by means of a cord and pulley, and let down as often

as

as an observation was to be taken. His upper Thermometer was always exposed to the sun; the two at 5 feet high, one in the sun and one in the shade. The result of the observations, the author tells us, was briefly this,† That in all seasons of the year, when the weather was serene, the Thermometer at the lowest station, in the greatest heat of the day, rose two degrees higher on the scale of REAUMUR,* than that at the upper station; and in the greatest cold of the night, de-

† Le Matin, environ deux heures ou deux heures et demie apres le lever du ſoleil, ces deux Thermomètres ſont d'accord, et indiquent une meme temperature aux petites oſcillations pres produites par des circonſtances accidentelles. A meſure que le ſoleil ſ'élève davantage le Thermomètre à 5 pieds de terre ſ' echauffe relativement à celui à 75 pieds; et leur plus grande différence qui a lieu dans le moment le plus chaud du jour, eſt d'environs 2 degrès de diviſion en 80 parties, dont le Thermomètre inferieur eſt plus élèvè que le ſuperieur. Le *maximum* de différence un fois paſſè, les deux Thermomètres ſe rapprochent, et quelque tems avant le coucher du ſoleil, ils ſ' atteignent de niveau, puis ils ſe depaſſent dans le ſens opposè; le Thermomètre inferieur ſe tiént plus bas que le ſuperieur, leur difference augmente rapidement dès que le ſoleil eſt couchè, et va juſqu' à 2 degrès, et quelqueſois davantage vers la fin du crepuſcule, Eſsais de Phyſique, par M. A. Pictet Tome 1. chap. 8. § 135.

* The ſcale of Reaumur contains 80 degrees between the boiling and freezing points.

scended

scended two degrees below it, and sometimes more.* What degree of excess the word *davantage* is here meant to express, I cannot precisely determine; probably nothing considerable; as a celebrated writer (M. du Luc) speaking of the difference of temperature observed by M. Pictet, between 5 and 50 feet with Thermometers, both exposed to the sun, gives the result in the same words; and then immediately draws this conclusion. †That in all seasons the sum of the heights of the Thermometer in one station, in the course of twenty four hours, is equal to the height in the

* The Thermometer at 5 feet in the shade, he says, from nine to three o'clock, the greatest heat of the day, agreed in its movement with those at 50 and 75 feet in the sun. The glass, therefore, at 5 feet in the shade, did not run so high as that in the sun, at the same time, by two degrees.

† Ainsi, d'après ces Expériences, la couche d'Air, distante seulement de 5 pieds du terrein, et celle qui en est à 50 pieds, sont à une même *Température* durant tout le cours des vingt-quatre heures, dans les tems uniformément couverts, les grand vents, et les Brouillards épais; et dans tous les autres cas, en toute saison, la somme des hauteurs du Thermomètre dans l'une durant les vingt-quatre heures, est égale à celle de ces hauteurs dans l'autre. D'où il résulte; qu'en tout, la couche d'air distante du terrein seulement de 5 pieds, n'éprouve pas sensiblement plus de *Chaleur*, que celle qui en est à 50 pieds.

De Luc, Idées sur la Météorologie—Tom. 2, p. 361.

 other;

other; that is, the mean temperature in the course of twenty-four hours was alike in both stations.§

The result of my observations, however, (see the foregoing table) will not admit of a similar conclusion. The excess of cold in the night, at the lower station, is by them much greater than the excess of heat in the day, although they were both in the shade: and this was commonly the case, when the nocturnal variation at the lower station was considerable. And besides this difference wich took place between the upper and lower stations, in the course of twenty-four hours, I found a more general change accompany the annual revolutions of the seasons: for the excess of heat in the day time, which in the summer was found at the lower station, in the winter diminished almost to nothing, while the excess of cold in the night rather increased.

Hitherto I have been treating of the difference of temperature in the air, between 5 and 50 or 75 feet, and between 6 or 9 and 200 feet from the ground. But the variation is found to be

§ It should be observed that Mr. Pictet's lower Thermometer, a few feet from the ground, placed in the shade, stood 2 degrees lower than that in the sun, and consequently agreed with that on the top of the pole in the sun.

more constant and regular, as well as more considerable, as we approach nearer to the surface of the earth. Here the refrigeration almost constantly takes place, notwithstanding violent winds, thick fogs, and cloudy skies; although at *such times* in a proportionally less degree. And I have never found the glass on the ground, in the night, warmer, or even so warm as those above it; except when rain or snow has continued to fall during the whole night, be the general temperature of the air either warm or cold; or, which I have very rarely observed, when the night, though clear, has been uncommonly hot, without the least dew on the surface of the ground. In these very opposite and different states, the surface of the ground seems to have produced similar effects: in the one it was exceeding wet, and in the other exceeding dry.

To compare the difference of temperature which frequently takes place on the ground, with that at an elevation of 7 and 200 feet above it, the following observations were taken. The Thermometers I made use of on this occasion, were of the smaller size, described page 9: the upper one suspended on the north side of the Tower of Canterbury Cathedral, in the shade, as before; that at 7 feet elevation, on a post erected for the purpose in an open pasture ground; shaded from the sun in the greatest heat of the day, but not inclosed or skreened from the open air. The lowest Ther mometer was laid on the same pasture ground, in an open place, not shaded; neither are the day observations by this instrument entered in the table. The grass on the spot where it was laid, being about 6 or 7 inches long, was pressed down to an angle of about 10°; on this the Thermometer was laid, being insulated or separated from the grass by the space of about half an inch. I have been thus particular in noting this circumstance, because the refrigeration is found to differ, according to the surface on which it is produced. On green turf, where the grass is not an inch long, the diminution will be less on some nights, by 1, 2 or 3 degrees, than where the grass is longer, and according as the cold is more or less intense.

		Cold In the Night.			Heat In the Day		Difference of Cold In the Night.			Diff. of Heat in the Day
1791		Grass	Post	Tower	Post	Tower	Grass & Post	Post & Tower	Grass& Tower	Post & Tower
Oct.	13	33	41	43	56	55	8	2	10	1
	25	11	21	30	51	48½	10	9	19	2½
	30	32	42	43	47	46	10	1	11	1
Nov.	8	12	21	29	47	—	9	8	17	
	9	22	30	38	54	51	8	8	16	3
	13	28	36	41	47	51	8	5	13	+ 4
Dec.	10	17	24	26	34	37	7	2	9	+ 3
	12	6½	15½	17½	—	—	9	2	11	
	18	17	28	39	39	38	11	11	22	1
	24	20	27½	32	44½	44	7½	4½	12	0½
	25	17	26	30	46	44	9	4	13	2
1792										
Jan.	12	7	15	17	—	—	8	2	10	
	13	8	17	19	31	31	9	2	11	0
	14	0	12	17½	—	—	12	5½	17½	
	19	10	22	31	—	—	12	9	21	
March	3	29	38	39	—	—	9	1	10	
	6	26	34	36	51	49	8	2	10	2
	12	12½	25	29	22	35	12½	4	16½	+ 3
	23	25	34	37	55	53	9	3	12	2
April	7	23	31	35	58	57	8	4	12	1
	8	27	38	42	59	57	11	4	15	2
	9	25	36	43	66½	62	11	7	18	4½
	10	29	40	46½	55	61	11	6½	17½	+ 6
	11	28	38	45	70	67	10	7	17	3
	13	35	45	51	76	—	10	6	16	
	14	41	49	52	67	64	8	3	11	3
	15	34	45	46	56	67	11	1	12	+11
	25	31	41	41	60	56	10	0	10	4

By the observations in the foregoing table, the difference between the temperature on the ground, and at 7 feet above it, in the night, is greater than between 7 feet and 200 ; except in one instance, when it was equal : and at the same time the excess of heat in the day, at 7 feet, above that at 200 was inconsiderable. And further to shew how much the diminution of heat *in the night* increases as we approach nearer to the earth, I have added another set of observations, with Thermometers placed one on the ground, one 6 inches only above it, and the third at 7 feet.

		Night.			Difference.		
		Ground	6 Inch.	7 Feet.	Ground and 6 Inch.	Ground and 7 Feet	6 Inch. and 7 Feet
1793 Jan.	4	12	17½	22½	5½	10½	5
	7	12½	18	23	5½	10½	5
	19	11	17	23	6	12	6
	20	18	22½	27	4½	9	4½
	22	18	22	26	4	8	4
	26	13½	21½	27	8	13½	5½
	27	13½	18½	24	5	10½	5½
	29	28	32	34	4	6	2
Feb.	4	22	26½	32	4½	10	5½
	5	18½	23½	29	5	10½	5½
	6	22	27	30½	5	8½	3½
	8	24½	28	31	3½	6½	3
	20	17½	23	26½	5½	9	3½
	21	20	25	30	5	10	5

1792		Night			Difference.		
		Ground	6 Inch.	7 Feet	Ground and 6 Inch.	Ground and 7 Feet	6 Inch. and 7 Feet
June	10	34½	39	41	4½	6½	2
	12	38	42	43	4	5	1
	18	47	50	51	3	4	1
	22	34	40	41	6	7	1
	23	31½	39	41	7½	9½	2
	25	41	46	46	5	5	0
	26	41	48	50	7	9	2
	27	47½	52	52½	4½	5	0½
	28	46	51	52	5	6	1
	29	52½	56	57	3½	4½	1
	30	54	57	57	3	3	0
July	1	45	50	50	5	5	0
	2	40½	48	48	8	8	0
	3	39	47	47½	7½	8	0½

The nocturnal difference in this Table between the ground and 7 feet being rather less than the first, is partly owing to the latter observations being all taken in summer, although the weather was then remarkably cold for the season; and partly because the first are selected as the most considerable which happened in a longer period of time, while these are included in a shorter.

From the whole of these experiments, an idea may be formed, not only of the diurnal, but also of the annual variation of temperature in the air at different altitudes, from the surface of the ground, to within the compass of these inquiries. This information may be useful, not only in directing us how to place Thermometers for making observations; but may also tend to elucidate some points of natural Philosophy, and to rectify some errors, which have inadvertently been introduced into it. An instance of this, if I mistake not, occurs in JEFFERSON's *notes on the state of Virginia*, where the author says, "*The access of frost in the Autumn, and its recess in the Spring, do not seem to depend merely on the degree of cold; much less on the air's being at the freezing point. White frosts are frequent when the Thermometer is at* 47°, *have killed young plants of Indian corn at* 48° *and have been known at* 54°. *Black frosts and even ice, have been produced at* 38°½, *which is* 6½ *above the freezing point.*"

I shall here observe, from Dr. BLAGDEN's experiments, that, under certain circumstances, water may be cooled several degrees below the freezing point, before congelation takes place; but it never freezes in a less degree, than about the freezing point of the Thermometer.

That

That white frosts should be seen on the ground when the Thermometer stood at 54°, is certainly, at first sight, a surprising phenomenon; but I flatter myself that, by the help of my Thermometer, I can fully account for it, without having recourse to a supposition that water will freeze in a less degree of cold, than that marked by the freezing point of the Thermometer.

At seven, or between seven and eight o'clock in the morning, I have seen the mercury in a Thermometer at 7 and 9 feet from the ground stand 11° higher than what the index of the same instrument marked as the greatest cold of the night: while the index of another at 1 inch from the ground, pointed still 11° below that at 7 feet. So that at seven and eight o'clock the mercury in a Thermometer at 7 feet from the ground, stood 22° higher than the greatest cold of the night on the surface of the earth. It was still more common to see the mercury at the same hour in the morning at 44°, when the nocturnal cold near the ground had been below freezing; and I have seen it at 38°, when the index of the Thermometer near the ground marked for the cold of the night 27°, or 5° colder than is sufficient to produce, in one night, ice of a considerable thickness.

This

This author's mistake certainly arises from not being aware of the nocturnal refrigeration which takes place on the surface of the ground where the frost appears; and as the range of the Thermometer is said to be greater in America than in Europe, so likewise this refrigeration may probably be greater than in England.

Again, " *That other circumstances must be combined with the cold to produce frost, is evident from this also, that on the higher parts of mountains, where it is absolutely colder than in the plains on which they stand, frosts do not appear so early by a considerable space of time in autumn, and go off sooner in the spring, than in the plains. I have known frosts so severe as to kill the hiccory trees round about Monticello, and yet not injure the tender fruit blossoms, then in bloom on the top, and higher parts of the mountain.*"

Here again the author's ideas respecting the cause of these phenomena, are erroneous as before, from not knowing, or not considering the effect of the nocturnal refrigeration on the ground. 'Tis true, on a hill of moderate height, the air in the day time is colder than on the plains, and even the mean temperature in the course of twenty-four hours is likewise so: yet the range of the Thermometer will be less, and particularly the nocturnal

nocturnal cold, which produces the sharp frost so injurious to vegetables. This is principally owing to the much smaller quantity of dew which is found on the mountains, than in the plains; and Mr. Jefferson is certainly right when he says, this privilege against the frost is undoubtedly combined with the want of dew on the mountains, which from long experience he had found to be very rare. For it is a well known fact, that frost will sometimes destroy vegetables in the plains, when on the hills they are unhurt; and that, because in the night, it is really warmer; which has been proved by thermometrical observations, when the frost in the plains, and its absence on the hills have exactly agreed with the temperature marked by the Thermometer: and Mr. J. had he made corresponding observations at the different stations, with Thermometers placed near the ground, and examined them at proper times, would undoubtedly have been convinced that frost appears uniformly in consequence of that certain degree of cold, marked by the freezing point of the Thermometer, and no other.

CHAP. VI.

HOW BODIES THAT ARE OF A DIFFERENT TEMPERATURE FROM THE AIR, AND WHICH MAY HAPPEN TO BE NEAR A THERMOMETER, SUSPENDED AT A DISTANCE FROM THE EARTH, ARE LIKELY TO AFFECT IT: WITH SOME EXPERIMENTS TO DETERMINE MORE ACCURATELY THE ANNUAL MEAN TEMPERATURE OF DIFFERENT PLACES.

IT is obvious that walls or buildings heated by fire, or other causes from within, or the sun's rays from without, must sensibly affect a Thermometer suspended against them. This circumstance is therefore carefully to be avoided; especially as such large bodies retain the temperature they have once acquired, a considerable time after that of the surrounding air is changed. It is best in all cases not to let the Thermometer frame come in contact with the wall, by the space of an inch

inch or more, that the air may thereby have free circulation round it, and prevent the instrument from being affected by it. I would prefer to all others, an open, shady, northern exposure. 'Tis true diaries are sometimes kept with Thermometers placed within doors, in some part of a building, where the temperature is supposed to be most equal: but as the temperature within different buildings depends on so great a variety of local circumstances, I must confess I think observations made in that manner are not so proper to shew the general temperature of the air, nor so fit to be compared with those taken at other places.

But before I proceed farther, it may be proper to consider the method of calculating what is called the mean temperature of the air, by thermometrical observations. The general practice I believe is from the aggregate of the diurnal to take the monthly mean, and from the aggregate of the monthly to take the annual mean: but the diurnal mean, from which both the others are deduced, is not, like them, taken from the aggregate of several intermediate observations, in the course of twenty-four hours; but from the two extremes, the greatest and least number *only* of degrees within that period. A question may hence arise, whether

ther the mean number taken from the two extreme heights of the mercury in the course of twenty-four hours, will give the mean temperature, in all respects the same, as if it had been taken from the aggregate of a certain number of intermediate observations, made (we will suppose) at every hour within that period.* It is certain that the two extremes of the diurnal temperature within the month, will not give the same monthly mean, nor the two extremes of the monthly temperature within the year, the same annual mean, as if deduced from the several intermediate observations. The question therefore respecting the calculation of the diurnal mean from the two extreme numbers, can be determined by experiment only.†

The most convenient method I could devise of obtaining information in this matter, was by placing a Thermometer so as to be slowly, but regularly affected by change of temperature: the motion of this glass was consequently regulated

* On this principle, I presume, obſervations are taken with Thermometers placed within doors.

† The heat does not increaſe and diminiſh, in reſpect to time, in exact arithmetical progreſſion.

as

as well by duration as by the degree of heat.‡ For this purpose I had one of my Thermometers inclosed in a block of Fir, so as to be closely covered by the wood, by about the thickness of two inches; this I suspended in the shade near to one by which my diary was kept. This inclosed Thermometer, though sensibly affected by the diurnal variation, had nevertheless its range considerably diminished; and when the variation by the glass in the open air was greatest, the indexes in this did not reach either of the extremes by 6 or 7 degrees. On comparing the mean of the diurnal observations by the inclosed Thermometer, with those taken in the open air, I found them sometimes exactly agree, at other times differ a half or a whole degree, inclining first one way, and then the other, according as the weather was in general inclined to vary towards heat or cold: but on comparing observations made in this manner, both in winter and summer, I found upon the whole a difference of one degree only, which in this case may be considered as nothing.

Another experiment of this kind, relating to the annual mean temperature, I made in the following

‡ And here it feems neceffary to know how long the temperature continues at any certain degree.

lowing manner. In a spot of ground, almost constantly shaded, I sunk a tube of wood, three feet long, nearly its whole length into the ground. The inner diameter was adapted just to receive one of my Thermometers; and observing at what depth the diurnal variation ceased to affect it, I fixed it there. This brought the middle of the glass two feet below the surface of the ground. Besides a plug of wood made closely to fill the upper end of the tube, about seven inches long, a box of wood likewise covered that part of it which rose above the ground. This I observed once a day. If the weather happened not to change so as to affect the mean diurnal temperature of the open air for a continuance, the mercury in the inclosed glass remained motionless; and in cases of continuance, after a considerable change in the temperature of the air, it hardly ever followed the change faster than one degree per day.

With Thermometers placed in this manner, I kept a regular journal for the space of two years; and the result was in a great measure such as one might reasonably expect. For as large bodies are slowly affected by a change in the air which surrounds them, so I found the temperature of the earth following the general change of the

atmosphere,

atmosphere, and in a great measure correspond-ing with its own annual motion: for after the month of March, when the sun gets above the equator, I found the monthly mean of the air greater by a degree or two, than that of the earth; and after the month of September, when the sun is below it, the heat of the earth was as much greater than that of the air. But notwithstanding this difference in the monthly, on casting up the annual mean of the air, and comparing it with that in the ground, it appeared that in the year 1790 there was no difference at all, and in 1789 no more than 0°. 02; so remarkably did the mean temperature taken in the ground agree with that taken in the air, although the stations of the two Thermometers were so very different.

By the result of these experiments the truth of the annual mean temperature of Canterbury, as before given, is farther confirmed; the mean temperature of the ground agreeing with it. But had the former been taken with other Thermometers, and in the usual manner, there would have been a difference of 3 or 4 degrees.

From the experiments made with the Thermometer in the ground so nearly agreeing with those made with that inclosed in the case of wood, and

 both

both together agreeing with those made in the open air, we may conclude, that although a small irregularity will sometimes happen in the diurnal observations, which upon the whole correct each other; yet no error of importance will arise from calculating the annual and monthly mean temperature in the usual way; that is, from the two extremes of the diurnal variation. It is likewise observable from the foregoing experiment, that the mean temperature of the earth is regulated by that of the air. Similar experiments may be made with common Thermometers in every soil and climate, by observing at what depth from the surface of the ground the diurnal variation ceases to operate; but it is necessary on this occasion, that the bulb of the Thermometer be covered to some thickness, to prevent the external air from affecting it at the time the observations are made.

CHAP. VII.

SOME EXPERIMENTS TO DETERMINE THE EFFECT OF PLACING A THERMOMETER IN THE SUN, OR IN THE SHADE.

THE heat which is produced by the sun's rays on different bodies varies considerably according to their size and the figure of their surface, as well as the direction in which they fall upon them: it depends still more on the texture of their component parts. How far chymical affinity, or other causes still more occult, may operate in this case, is not the subject of my present inquiry: but this is certain, that all dense bodies, especially those of the metallic kind, are greatly disposed to receive and retain the heat of the sun's rays; and the more so if their surface be not polished, as they are then less capable of throwing them off by reflection. The calorific power of the sun's sun's rays is almost imperceptible on such bodies as being extremely rarefied, as well as transparent, freely admit their passing through them. When they fall obliquely on the surface of bodies, especially such as are highly polished, their heating power decreases with the angles of incidence. Hence it should seem that the rays of the sun, without meeting with some dense opaque body, will not produce sensible heat.

It is to our present purpose to observe that the bulb of a mercurial Thermometer, although very dense, is, on account of its spherical figure and polished surface, very little liable to be affected by the sun's rays; which has induced an eminent writer to conclude that the rays of the sun do not sensibly heat the insulated bulb of a mercurial Thermometer.

To satisfy myself in this particular, on the 16th of July 1793, I made the following experiment. I suspended two mercurial Thermometers, in an open pasture, 7 feet from the ground; their bulbs and great part of their stems were perfectly insulated, and not at all skreened from the direct rays of the sun, or affected by any reflection of heat from other bodies. They exactly agreed with each other in their scales, and in their movement. At

12 o'clock

12 o'clock these glasses, the sun shining full upon them, stood at 89°½. I skreened one of them from the sun, by placing before it a slip of deal, one inch broad, and half an inch thick. In about four minutes time it had fallen four degrees, and stood at 86°½, whilst the other, which remained unshaded, had risen half a degree, and stood at 90°. In this state they both remained some time stationary. I then removed the shade to the other glass, which stood in the sun at 90°. and letting it remain as before, about 4 minutes, I saw this likewise had sunk, and stood at 86° ; whilst the first glass, from which the shade had been removed, had now risen to 90°. I first made the experiment placing the deal shade at five inches from the Thermometer; I afterwards repeated it at 12 inches distance, and the mercury sunk just the same as before. It is a circumstance worthy of note, that the wind blew briskly all the time this experiment was making.

When the general temperature of the air was about 80°, I have found the glass in the sun to differ from that in shade about four degrees; if the temperature was 70°. the difference was three degrees, if 60°-two degrees, or nearly so; for it varied a little according as the air was more or less loaded with vapour. I have for a month together kept

 half

half a dozen Thermometers constantly exposed to the sun, and as many shaded, in the manner above described; which in clear weather always differed in the extreme heights of the mercury, in about the same proportion as those just mentioned. Mr. PICTET also, placing two Thermometers at the same height from the ground, one in the sun, the other in the shade, found the former in the greatest heat of the day, rise 1 or 2 degrees higher than the other; which agrees with the result of of my experiment, where the difference was from 2 to 4 degrees of FARHENHEIT.

CHAP.

CHAP. VIII.

DESCRIPTION OF A THERMOMETER FOR TAKING THE TEMPERATURE OF THE SEA, OR WATER AT GREAT DEPTHS.

BEFORE I describe this Thermometer it may be proper to give a brief account of the different methods that have been already made use of for the same purpose.

The best authenticated experiments of the kind that I have met with, are those made by Captain DOUGLAS at sea in 1769. By Lord MULGRAVE in his voyage to the north in 1773. Those described by Mr. FORSTER in his voyage with Captain COOK in 1772 and 1773. And those by Mr. de SAUSSURE in the lake of Geneva in 1779.

Captain DOUGLAS made use of a tin cylinder containing a large quart, water tight, with a Thermometer fixed in it; this was let down, and left at the greatest depth for half an hour. After finding the inconvenience of making the experiment in this manner, because of the length of time it required, he made a small hole in each end of the cylinder, and let it down without first filling it with water. It sunk 260 fathom in 3½ minutes and it was hauled up in 13½.

In Lord MULGRAVE's voyage, a bottle properly prepared by Dr. IRVING, for bringing up water from the bottom without changing its temperature, was let down, fastened to the line about two fathom from the lead. A Thermometer was plunged into the water it contained as soon as it was drawn up. The bottle had a coating of wool three inches thick, which was wrapped up in oil skin, and let into a leather purse, and the whole inclosed in a well pitched canvass bag, firmly tied to the mouth of the bottle, so that not a drop of water could penetrate to its surface. A bit of lead shaped like a cone, with its base downwards, and a cord fixed to the small end, was put into the bottle, and a piece of valve leather, with half a dozen slips of bladder

bladder were strung on the cord, which when pulled, effectually corked the bottle in the inside.

A Thermometer by Lord CAVENDISH, constructed to shew the temperature of the water at the bottom of the sea, was also let down; but the result of the observations taken by the two different methods not perfectly agreeing, rendered both in some measure doubtful.

The Thermometer made use of by Mr. FORSTER was of FARHENHEIT's construction, made by RAMSDEN, and furnished with an ivory scale. It was on these occasions put into a cylindrical case, which had at each end a valve, admitting the water as long as the instrument was going down, and shutting when it was hauling up again.

Mr. DE SAUSSURE, in taking the temperature of the Lake of Geneva, tried the last mentioned method, with a variety of experiments of the same kind; but fixed on the following as the most likely means of accomplishing his purpose with accuracy. He inclosed a common Thermometer in a case of wood three inches thick, and also wrapped it several times round with cloth, so that it required to be immersed eight hours to bring it

to

to the temperature of water, which differed from it only 10 degrees. The Thermometer thus inclosed was left in the deepest part of the Lake of Geneva fourteen hours, and then drawn up in the space of ten minutes only. He afterwards fixed a Thermometer in a glass tube filled with water, inclosed in a case of wood and one of iron. With this also he made experiments in several other lakes, leaving it in the water sometimes a whole night for each experiment.

These observations of Mr. De Saussure were made with an accuracy, which the advantage of greater leisure had given him, and are undoubtedly correct; but the length of time required by this method, for making a single observation, would be too considerable to be generally practised at sea.

This ingenious writer, in consequence of his own experience, as well as his knowledge of what had been already done, concludes his account by saying, "Nous repéterons ces expériences dans d'autres lièux & à differentes profondeurs; nous pensons même aller les tentér dans la mer; car celles que l'on a fait jusqu' à ce jour, sont absolument imparfaites." Now, if this be the case, the inaccuracy of the experiments must have arisen principally

Pl. III.

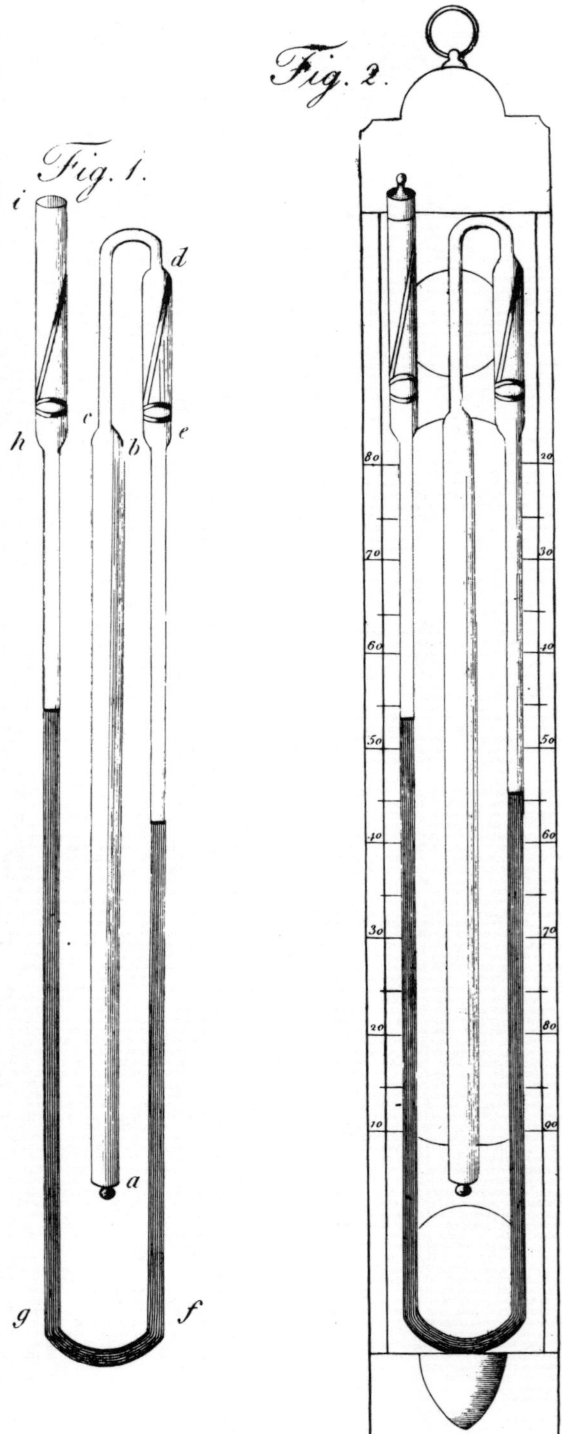

cipally from the defects of the instruments made use of on these occasions. If, therefore, a Thermometer can be contrived to ascertain more conveniently and accurately the temperature of the sea at a great depth, such an instrument may be useful: and with this view the following was constructed.

Pl. 3, fig. 1, *a b*, is a glass tube, nine inches long; the diameter of the bore $\frac{3}{8}$, the glass $\frac{1}{20}$ of an inch thick; *c d e f g h i*, a smaller tube, the diameter of which is varied in the following proportions. From *c* to *d* $\frac{1}{16}$, from *d* to *e* $\frac{1}{8}$, from *e* to *h* $\frac{1}{16}$, from *h* to *i* $\frac{1}{4}$ of an inch. The small tube is joined to the large one at *b*, carried up two inches above it, and down again on one side of it, in the direction *d e f*, and up again on the other side, in the direction *g h i*, in a plane with the center of the large tube, and at the distance of $\frac{1}{8}$ of an inch from its surface, and at the end *i* it is closed with a brass cap. This glass is filled with rectified spirit of wine to about an inch and a half from the end (or half the distance from *h* to *i*) except the lower part of the small tube, which is filled with mercury. Within the small tube on each side is an index, consisting of a piece of a needle or steel wire, an inch long, hermetically sealed in a very thin glass tube, just large enough to contain it.

When

When the instrument is not in use, these indexes or needles should, by means of a magnet, be both raised up and deposited in the receptacles *h i* and *c d*. When the instrument is to be used for observation, it should first be brought to the temperature of the air or surface of the water; and then both the needles be let down to rest on the surface of the mercury on each side. The instrument being thus rectified, it may be let down in the water to any depth that is required; and whenever it is affected by change of temperature, either increase or diminution of heat, the mercury on one side or the other will always descend, and the needle with it: but when it rises again the needle will not rise; but by remaining at the point of lowest descent, will shew the greatest degree of cold or heat to which it has been exposed after being rectified. On the scale on the side *e f*, which descends by increase of heat, the degrees are of course numbered downwards; and on the side *g h*, which ascends by heat, the degrees are numbered upwards. In forming the scale, they are ascertained by placing the instrument in a cold mixture with a mercurial Thermometer.

In respect to the construction of this instrument, it may be observed, the shape resembles that for shewing the temperature of the air in the observer's observer's absence; excepting the enlarged part of the tube from *d* to *e*, which is made for the purpose of lodging the index or needle, when the Thermometer is not in use, and thereby keeping it in a state ready to be rectified for observation. The glass also is thicker, to guard against the pressure of the water at great depths. The fluids are placed in the same manner as in the other; but the needles or indexes, a most essential part, act not only differently, but directly contrary to the former; these accomplishing by descending and remaining at their lowest point of descent, what the others do by ascending and remaining at the highest point. This Thermometer will also serve to shew the greatest variation of temperature in the air in the observer's absence, the same as the other; but being not so readily rectified, it is not so commodious for that purpose. This circumstance, however, renders it more proper for the use for which it is intended, *i. e.* shewing the temperature of the sea; because when the needles are immersed in the mercury, they are so forcibly pressed against the Thermometrical tube, that neither the most powerful magnet, nor the greatest shake of the instrument will remove them. To rectify it when the indexes are immersed in the mercury, it is necessary to bring it to the same temperature it was in when the needle took its

place.

place. If increase of heat should be necessary on this occasion, it may readily be done; but if a diminution is required, that may not so conveniently be obtained. But strong spirit of wine, or Sp. Vitr. D. or Ether, applied to the middle tube, will soon sink the mercury 18 or 20 degrees, by which means the needle may be set free, and lodged in its receptacle *h i*.

THE

APPENDIX,

CONTAINING

Directions for making Thermometers,

WHICH SHEW THE EXTREME VARIATIONS OF

TEMPERATURE IN THE ATMOSPHERE,

During the Observer's Absence.

THE minuteness of the following directions for the construction of these Thermometers may perhaps to some persons appear censurable, till they are reminded that twelve years have elapsed since an account of them was first published in the Philosophical Transactions; during which time imitations of them have been made in London, but so essentially defective in their construction, as to have brought disgrace rather than credit on the invention. This circumstance, together with the advice of many of the Author's Philosophical Friends, will, it is presumed, sufficiently justify their insertion, and clear them from the charge of insignificancy which might otherwise be brought against them.

Pl. IV.

Fig. 1.

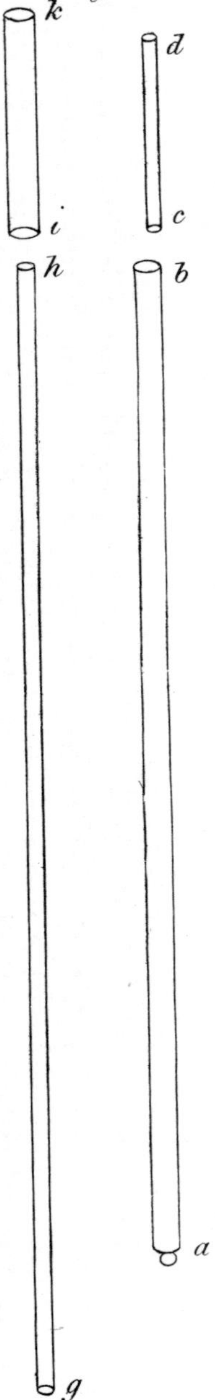

Fig. 2

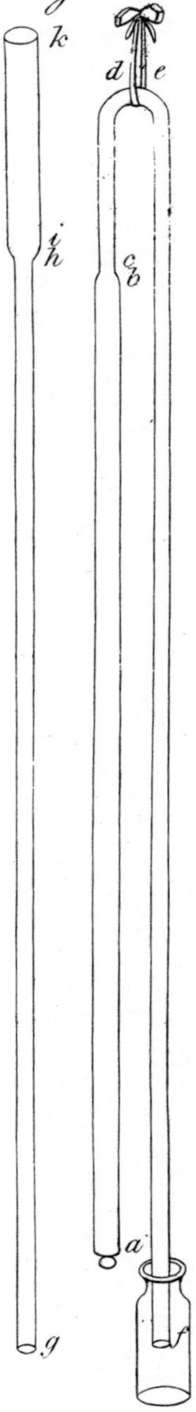

Fig. 3.

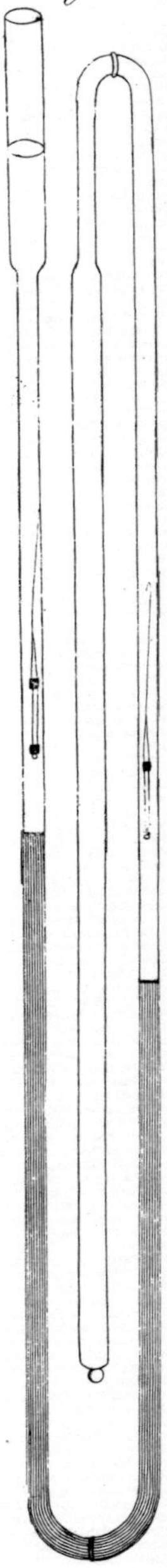

THE APPENDIX.

I.

How to select the most proper Tubes.

THE first requisite is, the choice of the Tubes, and the proportions they must bear to each other.

It is most convenient that they should be had from the glass-house in pieces as long as possible; because thereby a better opportunity is given of selecting such parts as are fittest for the purpose; which require to be as *strait*, *round*, and *even*, as can be procured.

The larger Tube, which is intended for the body of the Thermometer, should be very thin; that the spirit which it contains may be the sooner affected by any change of temperature. It should likewise

likewise be strait: but the inner diameter, or bore, in this does not require to be so perfectly equal throughout, as in the smaller tubes; provided it contains in the whole, the same quantity of spirit as one of a proper size that is perfectly even.

In the smaller Tubes, which form the neck of the Thermometer, it is proper the glass should be thick enough to give them strength and firmness; with the bore truly round, and as even throughout as possible: for wherever it is enlarged or diminished the pressure or force of the index-spring will be proportionally diminished or increased; and, if that should happen beyond a certain degree, the index in one case might not keep its place when the mercury descended; or in the other, it might be too stiff to float on it as it ascended. To guard against this inconvenience, and allow of a little irregularity in the tubes (which can scarcely be avoided) it is best to make the index-spring about an inch and a half long. It is necessary to observe also, that there be no grits in the glass, which appear like small grains of sand; for wherever they appear, the bore of the tube is certainly too much contracted to be fit for use.

For measuring or gaging the bore of the smaller tubes, I have made use of a piece of brass wire three

three inches long, $\frac{5}{40}$ in diameter at one end, and tapering to $\frac{1}{40}$ at the other; and divided in its whole length into ten equal parts. For the larger tubes, I have one of a similar form; $\frac{5}{10}$ of an inch in diameter at one end, and tapering to $\frac{1}{10}$ at the other; divided also in its length into ten equal parts. If this be of ivory and turned hollow, it may serve as a case to the other. These I have found extremely useful; for, by putting them into the ends of their respective tubes, the diameter of the bore is seen immediately with great precision; as also whether they be round or not; and, since the proportion I find best to be observed between the diameter of the body and the neck of these Thermometers, is that of four to one (whether the instrument be large or small, long or short); these simple tools shew at once, which of the tubes are most suitable to each other.

To ascertain the variations in the intermediate parts of the tubes in their first and longest state, I have found it convenient to wrap a slip of paper two or three times round the tube; which, upon being passed along from one end to the other, shews by the feel where the tube increases or diminishes in its size, and how far it is nearly equal; and that with sufficient accuracy to direct where to cut into shorter lengths such parts as are fittest for use. But the passing of a small quantity of mercury through the tube, in the usual way, certainly serves to ascertain the truth of the bore itself with greater precision.

It is extremely necessary that the neck of the Thermometer on one side should perfectly correspond with that on the other; which they will not do, if they are originally in one piece; and what little variation in size there may be at their different ends, should be so placed, as for the smallest part to be uppermost in both. That side of the neck which is to be put on last, must have a bason or air-vessel formed, or ready to be joined to its upper extremity.

II.

The manner of making the Indexes.

This is a very essential part of these Thermometers, which requires great care.

Provide yourself with some of the finest sewing needles; nip off $\frac{1}{3}$ of their length from the end which has the eye, and keep the remainder for use.

Make

Make some small glass tubes, such as may just fit the needles. This is to be done by taking a piece of tube, of that sort of which the body of the Thermometer is made (a thicker will not do) and bringing it to an even red heat quite round, for about the space of $\frac{1}{2}$ or $\frac{3}{4}$ of an inch: then, withdrawing it from the lamp, draw it out, but not too suddenly. In this way, small tubes of various sizes may be obtained; out of which, such as are round and of proper dimensions may be selected and cut for use.

Provide likewise a parcel of small black bugles. Select from among them, such as are suitable to your purpose; which may be done in the following manner.

Spreading some of them on a sheet of white paper, pick them up singly on a slender piece of Bamboo cane, cut very gradually tapering. As they are found to go farther or less on the Bamboo, their internal diameters will be known.

For ascertaining their external diameter, it will be convenient to cut half a dozen pieces of tube (of the sort of which the necks of the Thermometers are made) of about 1 $\frac{1}{2}$ inch in length; whose bores are true from end to end, but gradually differing from each other in size. These being set close to each other on a slip of wood, in the order of their bore, will serve for comparing the size of the bugles; which, while each is on the point of the bamboo, may be introduced into these tubes; when it will be seen to which they are best suited. The bugles may then be kept in separate little boxes, according to their sizes, and distinguished by different numbers.

Having now your materials properly in order, take a fine tube that will just admit one of the needles; and observing which part of the tube fits it best, scratch it there with the edge of a fine watchmaker's file, or the edge of a knife previously rubbed on a rough stone: for if the tube be not first scratched all round with an instrument of this sort, it will break obliquely; a circumstance which ought carefully to be avoided. Present then the end of the tube just broken to the lower part of the flame of a wax taper, because that part burns without wavering. The glass will immediately melt; and the tube will be sealed hermetically, terminating in a small knob; the diameter of which will not at first be quite so large as the tube itself, if it is not put too far into the flame. Having now some bugles of the proper size on a piece of white paper, take up two

two of them on that end of the tube which is sealed, making choice of such as just go freely on. Then, removing the bugles to a little distance from the end, put that again into the flame of the candle a little farther than before; and, giving the tube a rotatory motion between the fingers, to keep the melted metal from dropping, a knob will be formed in a line with the tube, and a little larger than the tube itself. The use of the knob is not solely to prevent the bugle from slipping off the end; for it is of great service, in causing the index to float readily on the surface of the rising mercury.

As soon as this knob is rightly formed, the tube must be withdrawn from the flame. The bugles may then both of them be brought down close to the knob on the outside, as also the broken end of the needle within: after which, the tube must be scratched as before, and broken off $\frac{1}{4}$ of an inch beyond the pointed or upper end of the needle. Then, holding this, which may be called the body of the index, between the thumb and finger of one hand, take in the other a small tube, just large enough to go upon the upper end of the body, and pass it so far as to coincide with the point of the needle. The end of this outer tube must be perfectly even before it is put on the other. Then, while the tubes are thus connected together, bring the part where they are double (but a little short of the end of the outer tube) to the flame of the candle; and, when they are melted together, removing them from the flame, draw out the spring, which should be done with a moderately quick motion. The end of the outer tube thus left, will serve as a shoulder to secure the upper bugle from slipping off. The spring may be nipped to a proper length (about $1\frac{1}{2}$ inch I think the best) and being presented to the flame, will be sealed with a small spherical knob, which may best be suffered to incline to one side.

Nothing now remains but to confine the bugles one to each end of the body, and to set the spring to a proper inclination; both of which must be done with care, suitable to the brittle nature of the materials. When the indexes are thus far advanced, they are best taken up, and handled as far as may be, with a magnet.

For confining the bugles, the method I have practised, which has scarcely ever failed me, is this: after they are put in their proper places, twine a single thread of the finest silk three or four times round the body, close to the bugle; and,

and, tying it fast, clip off the ends with scissars. Then singe them close, by just presenting them to the flame of a small taper. In doing this a steady hand is required, lest the spring be injured by too near an approach to the flame.

In setting the spring to a proper inclination, a little experience will enable a person to judge pretty nearly what will be required for tubes of a certain size. Yet this can never be determined with sufficient precision, but by experiment: *i. e.* putting each index into its own tube filled with spirit, and trying the force of the spring by a magnet, and adjusting it accordingly. This is to be done by presenting the lower part of the spring near to where it is connected with the body, towards the flame of a very small taper. The glass will soften before it comes quite into contact with the lower part of the flame, and in that state it may be set to any inclination whatever. In doing this, the inclination of the spring should be so turned, as for the knob at its end to be directed from that side of the tube against which the spring is intended to press.

Before an index is put into its place, it may be adviseable to soak it and wash it well in warm water, lest it should have contracted any foulness from

from the flame, or handling, which it would communicate to the spirit.

III.

How to join the first portion of the Neck to the Body.

All things being thus prepared, let one end, *a*, (pl. 4, fig. 1.) of the body, *a b*, be hermetically sealed; and to the other end, *b*, let the small connecting piece, *c d*, be joined. Tie them now on a slip or frame of wood, two inches shorter than the body itself; so that the body may extend about an inch beyond the wood at each end.

Join then what is intended for the upper end of the first part of the neck, *e f*, at *e*, to the connecting piece at *d*; and, making the connecting piece flexible, bend it round so as for the neck-piece to be brought parallel to the body; in doing which, the frame on which it is fixed will be a guide. It may then be taken off the frame, being ready for fillng.

IV.

How to fill this first part and put in the Index for Cold.

Fill a two ounce phial nearly full, with spirit of wine, highly rectified, so as to fire gunpowder.

Let

Let the cork of the phial be perforated, so as just to admit the lower end *f*, fig. 2, of the tube *e f* into the phial, as far as it descends below the end of the body. Let a piece of string be tied about the neck of the phial, and after slightly twining round the tube *e f*, extend above it, so as to be held in the hand; together with another piece of string looped on the upper curvature of the Thermometer. In this way both may be supported together by the same hand, at any convenient distance above them.

Suspended thus, let the body be presented to the front of a stove or coal fire, in such a manner, as that the phial may hang below the fire, or in some way be screened from it. The air by its expansion will be driven by the heat of the fire from the body of the Thermometer into the phial, and will rapidly pass the spirit in bubbles. When the current of air begins to cease, remove the whole from the fire. As it cools, the spirit will ascend the neck, and a small part of it will pass over into the body. Heat it again in the same manner, almost to a degree of boiling; and remove it to cool as before. After a repetition or two the body will become nearly full. When that is effected, it must then be heated just enough

to

to make the spirit boil; which will for a moment drive a great part of it back again into the phial: but upon removing it from the fire, the spirit will suddenly return and fill the whole body; excepting a small bubble of air which will remain, about the size of a pea. If too much air be left, it must be heated again till the quantity is diminished.

Hang it now in as cool a place as can be found; and when it is quite cold, disengage the Thermometer from the phial.

In this stage warm it again; and after the air-bubble which rested at the top of the body, is by the expansion of the spirit driven through the curvature into the side of the neck, invert the whole, and put down into the smaller tube, so far as just to reach the air-bubble, an exceedingly slender piece of Bamboo cane, by the side of which the air will ascend, and leave the neck filled to a certain height with spirit only.

The body and part of the neck of the Thermometer being now filled, one of the indexes must be put in with its spring foremost.

How

V.

How to join on the remainder, and fill the whole; as also how to handle the tubes when detached from their frame.

Join on next the remaining parts; *i. e.* the second half of the neck, *g h*, and the air-vessel, *i k*, at its extremity. Then, tying the Thermometer again on the little frame above-mentioned, render the tubes flexible by the flame of a lamp, and set them perfectly parallel to the body. See Fig. 3.

While it is still on this frame, it may be advisable to loop a piece of string round the upper curvature, *d e*; and then, detaching the glass from the flame while they both lie in a horizontal position, to raise them together into a vertical one; after which, taking the glass carefully off the frame by means of the string, it may be applied to another frame.

Another method of handling the tubes in this very tender state, is, to have two pair of common cards lying upon each other, and pinned together at one end each pair; by which, after slipping one card of each pair between the tubes and the frame, at sufficient distances from each other, you

may

may inclose the three tubes between the cards, and then, keeping them together, transfer the whole with safety.

The Thermometer being, in either of these ways, or in any other that may be found more convenient, taken off this frame on which it was made, must now be put on another, pierced down the middle so as to leave the body insulated. Warm it now before a fire so as to cause the spirit to pass into that part of the neck, *g h*, which was last put on; and, as soon as it has passed the lower curvature, filling the air-vessel with spirit, put down the slender Bamboo far enough to reach the spirit at the bottom. The spirit from above will immediately descend and join that below; and the air, ascending in bubbles by the side of the Bamboo, will escape, leaving the whole of the tubes now completely filled with spirit.

VI.

How to put in the Mercury, and pass it to its proper place.

The Thermometer is now ready to receive the mercury. For this purpose, let it be heated ten or or fifteen degrees hotter than the greatest degree of heat you intend for your scale. While it is in this state, fill the whole side, *g k*, with mercury, passing down the Bamboo to a certain point near the bottom; by the side of which the mercury will descend through the spirit just to that point and no farther. Remove it then into a cooler place; and, as the spirit contracts, the mercury will follow it round the curvature, and rise in the opposite side, *e f*. Compare it now with another Thermometer; and observe, to what points the mercury rises on that side by certain degrees of cold, as also what space is occupied by any number of degrees. This will enable you to form a judgement of the range of your scale. If the mercury should not rise high enough, (and it is best not to carry the mercury, by the first heating, quite so far as may be finally intended) heat it then just so many degrees hotter than you did before, as you wish to have the mercury advanced; then passing down the Bamboo exactly to the same point as you did before, the mercury will descend thither; and when all is cool, it will rise just so much farther on the other side. Thus, by trial, and by comparison with another Thermometer, it may be ascertained to one degree, how much it requires to be advanced, and may be brought exactly to that.

When

When this end of the column of mercury is adjusted, that on the other side is easily brought to what point you please by an addition of mercury. After which the index for heat must be put into its place, the body first, with the spring upwards; and the air-vessel be closed.

VII.

How to make the Scale.

The scale to these Thermometers must be done by comparison with a good standard mercurial one: immersing them wholly in a vessel of water, heated and cooled to the several degrees that are necessary.

Let a cylindrical vessel, of sufficient depth, (about two feet deep, and one in diameter) be filled with water, heated nearly to the degree first required. Let two other vessels be likewise at hand; one with very hot water, the other with water as cold as can be procured; by the addition of one or other of which, the water in the principal vessel may be brought to any degree of temperature that is not very cold. These additions it will be adviseable to lade by hand in some small vessel; since the quantity and proportion

portion will be thereby the better known. A flat circular piece of wood, of about 2 $\frac{1}{2}$ inches diameter, fixed like a churn-dash to the end of a small rod, will be necessary, to be frequently moved up and down in the water, to bring the whole to an uniform temperature. There should also be a cock near the bottom of the vessel, to let out superfluous water.

In this way the Thermometer should be tried successively, at every 5 degrees shewn on the mercurial one; which must remain in the water some minutes at each trial before it be examined, that its scale may acquire the proper degree of heat. If the vessel be large, and the temperature not differing too much from that of the room, the same degree of heat in the water will continue for a considerable time.

In marking the degrees above 40 or 50, it will be found convenient, that the air be somewhat cooler than the water; because the index on the side for heat, after having been raised to the proper point in the water, will be left there by the mercury, on taking the thermometer out of the vessel into the air, when the exact point at which it stands may be marked at leisure. In doing this, the lower edge of the lower bugle on the index is the point to be marked (and to be attended to in all observations with these instruments) and not the end of the knob of glass below it, nor even the surface of the mercury itself; whose two ends will be a little short of coincidence, in proportion to the buoyancy of the indexes.

In a similar way may those degrees which are below 40 or 50 be marked on the other scale, which is to shew the greatest degree of cold, provided the external air be warmer than the water in the vessel: the mercury, upon taking the instrument out of the vessel, leaving that index at the place to which it had been driven by the cold of the liquor, the bottom edge of whose lowest bugle must be marked on that scale.

To ascertain the freezing point and the degrees below that, it is necessary to provide a cold mixture. For this purpose, fill a cylindrical tin vessel (of about two feet high, and six inches in diameter) with pounded ice and sea salt; or, if ice is not to be had, with snow in its stead. The ice being put into a strong bag, should be beaten well with a mallet, till no piece of it is larger than the tip of one's finger.

The ice and salt should be dropped into the vessel by different hands at the same time, in the

the proportion (by weight) of two of ice to one of salt. When the vessel seems full, pour in two or three quarts of very cold water; and stir the whole with a stout stick till it becomes fluid enough to receive the Thermometers. Great care must be taken to keep it often moved; otherwise the temperature will not remain equal throughout: and the mercurial Thermometer must frequently be shifted to every part, to see that the temperature of the whole is alike.

This mixture, when it first becomes fit to receive the Thermometers, does not always produce exactly the same degree of cold. It is commonly about the Zero of FAHRENHEIT's scale; but sometimes a few degrees below it. As it would be extremely difficult to adjust the temperature of this mixture exactly to every fifth degree, it is best to take at first the greatest degree of cold that is obtained; and, as that diminishes (which it does not rapidly, especially if the air be as it should be below the freezing point) to mark on the frame at every three or four degrees, where the temperature seems at any time to remain a little while steady. If the mixture does not become warmer so fast as may be thought convenient (which is most commonly the case) a little addition of water will hasten it.

 The

The depth of the vessels here given, is for the Thermometers of the largest size. For smaller ones, they need not be so deep.

After the principal points of the scale are thus ascertained, and marked on those sides of the instrument where they will be most usually observed; they may respectively be transferred to their opposites, and the intermediate divisions may be marked on both sides the deal frame; whence the whole may readily be set off at any time on another. But before that is done, it will be best that the Thermometer be exposed in its present state to different temperatures to be satisfied that all is right, before it be considered as completed.

VIII.

How to repair any Accident to which these Thermometers may be liable.

1. In the first place, where the air vessel at the upper end is stopped with cork, or any other method but hermetically sealing, the spirit above the surface of the mercury on that side, will, after a length of time, diminish by evaporation; and the remainder becoming weaker, may be liable to

to freeze with any great degree of cold. On this account, if the spirit is seen to be much diminished, it will be adviseable, before the coldest weather of the winter comes on, to draw off that spirit, and replace it with some in a highly rectified state.

To effect this, take off carefully both the cap and base of the larger Thermometer; and, whether to the larger or smaller ones, take the stopper from the mouth of the air vessel; and, with a magnet, draw out the index on that side. Let down a slender piece of Bamboo so far into the tube as almost to touch the mercury. Then invert the tube, and the spirit will run out; especially if the Bamboo be a little moved with a rotatory motion to assist it, if that be found necessary. As soon as the spirit is all out, the instrument must be returned to its erect position, and fresh spirit poured in, which will run down freely by the side of the Bamboo to its proper place. Care must be taken not to let the Bamboo touch the mercury; because if it did, that would run out with the spirit. After this, replace the index, and close the tube as before.

2dly. It sometimes happens, especially in a Thermometer newly made, that a thin film will be seen on the surface of the descending mercury; which

which when that comes to the farthest point of its descent, will be left against the side of the tube, while the mercury will rise bright and clear above it. This, if it happens on that side which descends by increase of heat, is easily removed out of the common range of observation, by bringing it down below that range, and leaving it there. If it be on the other side, open the air-vessel and take out that index as before; then, putting down the Bamboo into the mercury till it reaches the film, by giving to the Bamboo a rotatory motion, the film will be disengaged from the sides of the glass, and may be drawn up into the spirit; which, becoming foul thereby, must then be taken out and replaced by fresh spirit, as was directed above.

3dly. If some of the mercury should at any time pass the index; let as much of it as possible repass by sinking that surface; and endeavour to remove what remains, by gently jarring the Thermometer on the knee, holding it in a vertical position. That will probably remove some: but if it should not remove all, place the Thermometer nearly in a horizontal position; and, with a very powerful magnet, force the index higher up. This will release some more; and then, by moving the index briskly, the rest of the mercury may be be broken into small particles, and thereupon all repass.

Yet is this but a temporary expedient; for, while the same causes continue to operate, the same effects will be apt to ensue. It will therefore be best to try to remove the cause.

If an index is ever obstructed in rising, it must either be, That there is some contraction in the tube at that place; or, That the force of the spring is too great; or, That something being got into the spirit, and sinking down, is wedged in between the bugles and the thermometer tube.

If it proceeds from the first cause, it is remediless; the glass must be broken.

If, from the force of the spring, the index must be taken out and altered.

If any small particle of dirt has got into the spirit: that cannot well happen, but on that side where the index can easily be taken out and cleaned; which it should be, otherwise the instrument cannot long be kept in order.

4thly. If at any time it should be found necessary to take out all the mercury: first opening the air-

air-vessel, take out the index; then passing a Bamboo down to the bottom of the tube, heat the instrument till all the mercury comes to that side. Invert it, and the mercury will directly run out: but its place must immediately be supplied with spirit, by returning the instrument, before it be at all cooled; otherwise the air will pass into the other side, and require a farther heating of the instrument to remove it. The mercury may afterwards be introduced as it was at the first.

5thly. If the Thermometer should by any accident be broken, either at the upper or lower curvature, it may be repaired, by putting in a short piece of tube, for as long as the body remains entire and filled, and the side tubes remain unbroken, they may be put together again, and the same scale will remain without error.

FINIS.